DIXIE LEGLER AND CAROL M. HIGHSMITH

Historic Bridges of Maryland

MARYLAND DEPARTMENT OF TRANSPORTATION

STATE HIGHWAY ADMINISTRATION

U.S. DEPARTMENT OF TRANSPORTATION

FEDERAL HIGHWAY ADMINISTRATION

MARYLAND HISTORICAL TRUST PRESS

Published in 2002 by the
Maryland Historical Trust Press
100 Community Place
Crownsville, Maryland 21032

Copyright © 2002 Maryland
State Highway Administration
707 North Calvert Street
Baltimore, Maryland 21202

All rights reserved, including
the right of reproduction in
whole or in part in any form.

ISBN 1-878399-80-2

Printed and bound in Singapore

Produced by Archetype Press, Inc.,
Washington, D.C.
Diane Maddex, Project Director
Gretchen Smith Mui, Editor
Carol Peters, Editorial Assistant
Robert L. Wiser, Designer

C O N T E N T S

BLACKWATER
RIVER

STATE ROAD COMMISSION

Blackwater River Bridge · ca. 1930s, Dorchester County

⬅️ This rustic wooden-beam bridge, no longer extant, is a type once found in abundance in the quiet, low-lying Tidewater region. As traffic increased, most rural timber bridges were replaced. Today a prestressed concrete bridge, built in 1991, takes Maryland Route 335 over the river.

Paper Mill Road Bridge · 1922, Hunt Valley, Baltimore County

⬆️ Designed by John E. Greiner, Paper Mill Road Bridge over Loch Raven Reservoir was modeled after the famous Hell Gate Bridge (1917) in New York City, the work of the noted bridge engineer Gustav Lindenthal (1850–1935). Greiner regarded his own bridge as a "daring and handsome structure . . . Lindenthalic in all its features."

Preface

We take bridges for granted in the early twenty-first century. Yet bridges have always been major achievements. They remove rivers as barriers. They move us forward on our journey. But historic bridges also allow us to look back in time, to appreciate where we have come from.

As a steward of Maryland's highway bridges, the State Highway Administration takes special precautions with historic bridges in its care. In the late 1990s, to develop a long-term preservation plan, SHA began working with the Maryland Historical Trust to inventory and document all of Maryland's historic highway bridges that are at least fifty years old. This book is an outgrowth of that effort.

Most of Maryland's historic bridges are on state or local roads, and those on state highways and roads are under SHA's stewardship. Some well-known historic bridges, however, were built for railroads or canals or have been taken out of highway use and converted to pedestrian or bicycle use in parks. Because they all contribute to the state's transportation history, the story of Maryland's historic bridges would not be complete without them.

Geography and local circumstances tend to divide the state into four general regions: the Eastern Shore, where navigation is critical and bridges must be movable; the central and southern region, where the railroad inspired new forms for the American bridge; the northern area, where metal trusses and covered bridges span rivers and streams; and the western region, where the nation's first federal road sliced through mountains and where stone bridges dot the countryside.

The story of Maryland's bridges begins with the early settlements that grew up around the Chesapeake Bay and the westward movement of hardy pioneers forging their way into the frontier wilderness. Rivers and streams were major impediments they attacked with resourcefulness. Location and native materials determined the type of bridge to be built. A deeply cut mountain stream called for an entirely different kind of bridge than did a slowly moving coastal inlet.

It is hard to overestimate the importance of a new bridge to Maryland's earliest settlers, when the only way for a farmer to get his cattle or grain across a river was to trudge right through the water. A new bridge changed everything. It also brought pride and prosperity to a community.

Today we can look back with admiration on local farmers who gathered hand tools to construct a covered wooden bridge, or masons who sliced great chunks of limestone to form into arches, or laborers who drove rivets into place for metal trusses, or crews who balanced hundreds of feet above the eddying Chesapeake Bay spinning cable for the giant suspension bridge. The stunning landmarks they left behind have a story to tell. Once merely practical, these historic bridges are beautiful windows to the past.

— ABBA LICHTENSTEIN, P.E., DR. ENG. (HC)

Introduction

Masemore Road Bridge · 1898, Hereford, Baltimore County

From wagons and buggies a century ago to cars and trucks today, this picturesque Pratt through truss is still moving traffic on Masemore Road over Gunpowder Falls. Located near the site of a 1700s stone mill and farm complex, the 119-foot-long bridge was built by the Wrought Iron Bridge Company of Canton, Ohio.

Thomas Viaduct · 1832–35, Elkridge, Howard County

This impressive eight-span, 612-foot-long stone arch railroad bridge, in service since 1835, pays tribute to the engineering genius of its designer, Benjamin Henry Latrobe Jr. The local topography required that the bridge make a difficult four-degree curve as it crossed the Patapsco River valley.

Echoes of a Simpler Time

Traveling along Maryland's back roads, past sturdy white barns and lush green cornfields, the winding path dips lower and lower, toward the stream. Down the hill, around the bend, and there it is stretching across the water: a bridge that has forded the centuries. Whether rugged limestone arches, delicate metal trusses, or protective covered spans, Maryland's historic bridges grace the landscape. And for a moment time stands still. The echo of horses' hooves and the rumble of wagon wheels seem almost palpable. The crack of a whip fills the air.

Historic bridges are essential features of Maryland's landscape. Their appearance along isolated country roads evokes a simpler time when common problems were solved with common sense. They moved goods to market and armies to the battlefront. Some have stood for nearly two centuries. Many are celebrated landmarks associated with the Civil War, the National Road (America's first federally funded road, also known as the Cumberland Road), or rail lines. Others are little-known spans that safely carried inhabitants and their families over raging streams. Together they present a fascinating portrait of early American ingenuity.

Sturdy bridges and good reliable roads were critical to Maryland's success as a colony and a state. But building bridges across Maryland's varied terrain has not been easy. Veined by a lacy network of rivers and streams, pierced by the jagged fingers of the Chesapeake Bay and the Potomac River, bounded on the west by the rugged Appalachian Mountains and on the east by the low-lying Atlantic seashore, Maryland's topography has presented incredible challenges. Fortunately, some of the nation's greatest bridge builders plied their trade here. Whether anonymous local craftsmen shaping stone and wood or highly trained engineers calculating stresses and loads, they built the bridges that bound Maryland together as a state.

Maryland's first highway was actually a waterway: the mighty Chesapeake Bay. Early settlements clustered around tidal inlets and navigable streams that emptied into the vast estuary. As the population grew and settlers moved inland, they created wagon roads from Native American trails and basic bridges from logs tossed across a stream. Ferries were established at prominent fords where crossings were deep. But as traffic increased, settlers responded with simple bridges of hewn logs set across upturned, forked tree trunks. Primitive and susceptible to floods, fire, and rot, they had to be continually replaced.

By the dawn of the nineteenth century, a spidery network of private turnpikes, or toll roads, reached out from Baltimore to meet the National Road at Cumberland and the burgeoning markets of the West. Early travelers overcame enormous odds, battling up and around Maryland's ridges and valleys with their wagons and mule teams. Reliable river crossings were critical, and local citizens demanded stone bridges that would last. Turnpike companies complied, hiring masons and laborers to construct dozens of limestone arches. Isolated homesteads and mill sites grew into full-blown communities as sturdy bridges made river crossings dependable year-round. Maryland's oldest extant bridge, the 1809 Parkton Stone Arch Bridge over Little Gunpowder Falls in Baltimore County, is a rare survivor of the turnpike era.

In the flat Tidewater region, where elevation drops are not so large, wood was still the best available material. Timber bridges here were built with short multiple spans, able to distribute the load as they pressed into the soft mud. Many were movable bridges with spans that could lift or swing out of the way to allow boats to pass.

The Susquehanna River drew to the region several nationally prominent bridge designers, including Theodore Burr (1771–1822), inventor of the Burr arch truss (an arch added to a triangular truss for strength). In 1817 he built an impressive eighteen-span, 4,170-foot-long covered wooden bridge over the Susquehanna River at Port Deposit. When that burned in 1823, it was rebuilt by Lewis Wernwag (1769–1843), a native of Germany who became a prominent American bridge designer. Wernwag also constructed a seven-span, 1,334-foot-long covered bridge over the Susquehanna at Conowingo that was later lost to a flood.

These giant covered bridges represented the high point of carpenter-built bridges. Most of Maryland's covered bridges, however, were not as ambitious. Many were one or two spans privately built by anonymous local farmers. Users often had to pay a toll and were urged to walk their horses through, as the vibrations caused by a galloping animal could jar supports loose.

Skilled carpenters, resourceful farmers, expert stonemasons, and brilliant designers made their mark bridging Maryland's growing road system. Their efforts soon paid off. By about 1825 Baltimore was the third-largest city in the United States, a major port, and the terminus of seven turnpikes. A variety of wooden bridges laid the groundwork for transporting goods in and out of the city, but competition with other ports was fierce. Three new transportation routes, designed to help Baltimore capture more of the southern and western trade, soon eclipsed the National Road and the state's turnpike system. The Chesapeake and Ohio Canal and the Baltimore and Ohio Railroad, both launched in 1828, and the Chesapeake and Delaware Canal of 1829 reduced turnpike travel to a trickle. But they also signaled momentous changes in bridge design.

The highly innovative Baltimore and Ohio Railroad, founded in Baltimore, was by far the most influential. The nation's first railroad, the B&O employed a virtual who's who of early American bridge designers, including such distinguished civil engineers as Stephen H. Long (1784–1864), Benjamin Henry Latrobe Jr. (1806–78), Wendel Bollman (1814–84), and John E. Greiner

Winter's Run Covered Bridge

ca. 1890, Van Bibber, Harford County

⬆ Covered only by a gabled roof, this wooden bridge over Winter's Run on the Philadelphia Road (now Maryland Route 7) reveals the structural beauty of the giant curved Burr arch that supported it. Its lack of protective walls no doubt contributed to its demise.

Havre de Grace Bridge

1907, 1927, Harford County

⬇ This toll bridge over the Susquehanna River was notoriously narrow and the second deck, added later, so low that large trucks became wedged. After the Hatem Bridge on U.S. Route 40 opened in 1940, the double-decker was demolished; only the piers remain.

29th Street Bridge · 1937, Baltimore City

Veneered in stone and edged in gleaming white concrete that resembles marble, this graceful closed-spandrel arch design is a testament to the art of bridge building. Designed by the J. E. Greiner Company, it consists of eight concrete arches: two central arches, each 232 feet long, bracketed by six 30-foot spans.

Calvert Street Bridge · ca. 1880, Baltimore City

One of Baltimore's most beautiful nineteenth-century bridges, the two-span bowstring metal arch on Calvert Street, designed by Charles H. Latrobe, no longer stands. When built, it was on the city's edge, which Baltimore's mayor, Latrobe's uncle Ferdinand, declared "a suitable place to promenade."

(1859–1942). The B&O introduced stone viaducts, the Long and Bollman trusses, and metal girders, demonstrating their safety and strength in carrying the heavy "iron horse."

The company built the nation's first stone railroad bridges, putting masonry to its most rigorous test yet. America's oldest railroad bridges, the Carrollton Viaduct (1829) in Baltimore and the Thomas Viaduct (1832–35) at Elkridge, both still in use today, attest to the durability of stone and to the extraordinary skills of these early designers and masons. But stone was heavy and expensive, so when cast and wrought iron became available, railroad engineers began exploring the possibility of using it to make bridges lighter and stronger and cheaper. Maryland became a laboratory for experimentation in adapting iron to railroad bridge design.

In a gradual intuitive process, engineers substituted iron for wood to improve rigidity and strength. Then in 1850 Wendel Bollman, the B&O's "Master of the Road," made a technological breakthrough with the first all-iron bridge to be widely used on the railroad. His iron combination truss-and-suspension bridge was a signal event in U.S. bridge building, heralding the railroad's commitment to cast-iron construction, which in turn led to its acceptance and use in Maryland's roadway bridges. The structurally redundant bridge worked so well that the B&O patented the design and used it widely.

In time engineers developed and patented a wide range of metal truss types—Howe, Pratt, Parker, Warren, Whipple—each named for its inventor and based on a different arrangement of triangular members. Built from cast iron, wrought iron, and later steel, they were designed and

prefabricated at a foundry by one of several bridge companies and then shipped to the building site in pieces to be assembled. County officials in charge of Maryland's roads were impressed with the new, sturdier metal trusses. Soon covered wooden spans, locally made, gave way to more reliable, longer-lasting, prefabricated metal bridges designed by professional engineers.

After the Civil War, as the technology for iron and steel making progressed, designers learned to shape metal into aesthetically pleasing arches. Metal arches were graceful, strong, and excellent for deep crossings with sturdy rock support. Baltimore was home to three spectacular large-scale examples designed in the 1870s and 1880s by Charles H. Latrobe (1833–1902), son of Benjamin Henry Latrobe Jr.: the Calvert Street two-span bowstring arch, the St. Paul Street through arch, and the Cedar Avenue deck arch, all now gone.

By the turn of the twentieth century, Maryland's deeply rutted and nearly impassable roads, long neglected in favor of the railroads and canals, slowly began to improve. By the 1920s the state's modernized roads surpassed rail as the primary means of moving people and goods. With an automobile revolution in full swing, state-sponsored road- and bridge-building programs proliferated, thanks to the state's Good Roads Movement and creation of the State Roads Commission in 1908. The landmark federal Good Roads Act of 1916, which released federal funds for road construction in anticipation of an interstate highway system, provided added incentive.

To meet the constantly increasing demands of automobile and truck traffic, bridges had to be stronger than ever. By the 1930s reinforced-concrete bridges and steel girders became the most popular new bridge types in the state. Most concrete bridges were standardized designs— beams, slabs, and rigid frames—but the malleable nature of concrete created a renewed interest in arch designs and ornament. As the structural possibilities of reinforced concrete became clear, arches grew wider, thinner, flatter, and more graceful. Eventually spandrel walls were opened into delicate ribs and columns and parapets were adorned in ornament, making concrete arch bridges the perfect choice for wooded, parklike settings.

In 1937, under the stewardship of Governor Harry W. Nice, Maryland launched the Primary Bridge Program to provide a continuous north-south highway connecting Philadelphia with Richmond, Virginia, while bypassing Washington, D.C., and Baltimore. The plan, ratified by Congress as part of its authority for navigable waterways, called for crossings over the Potomac, Susquehanna, and Patapsco Rivers and the Chesapeake Bay. The Potomac River bridge built under this program was named for Nice in honor of his leadership.

With the opening of both the Nice Memorial Bridge and the Hatem Memorial Bridge over the Susquehanna River in 1940 and the astounding 4.03-mile-long Chesapeake Bay Bridge in 1952, Maryland's engineers proved that they had the courage, muscle, and skill to span what was thought to be unspannable. The Hatem, a series of steel trusses, and the Nice, a combination truss-and-cantilever bridge, both comprise an intricate array of struts and angles. The Bay Bridge, joined by a second parallel suspension span in 1973, is supported by giant cables draped over massive towers. With their completion, once-remote areas on the Eastern Shore roared to life.

As these landmarks of American ingenuity show, there were few barriers that Maryland's bridge builders could not cross. These pioneers have bequeathed the state a great legacy. Charming and romantic, elegant and enduring, their bridges contain moving echoes of Maryland's past and powerful reminders of their creators' heroic achievements.

Kent Narrows Bridge

ca. 1930s, Kent Narrows, Queen Anne's County

⬆ Timber piles sunk into the bay floor support this single-leaf bascule, the first roadway crossing from Kent Island to the Eastern Shore. In 1950 it was replaced by another drawbridge, which still stands, and in 1990 a high girder bridge was built to carry U.S. Routes 50/301.

Williamsport Bridge

1909, Williamsport, Washington County

⬇ Williamsport Bridge, rehabilitated in 1949, 1980, and 1991, carries U.S. Route 11 over the Potomac River, Chesapeake and Ohio Canal, and Western Maryland Railroad rail lines. The addition of parallel plate girders has altered its original appearance.

Masonry Arch Bridges

The durability of stone and the strength of the arch make a powerful combination for a strong, long-lasting bridge. Using methods perfected by the ancient Romans more than two thousand years earlier, Maryland masons placed stones side by side over a wooden barrel-like brace and locked them into place with a keystone. Once the keystone was set, the falsework was removed. Rubble, large rocks, and dry soil filled in the space behind stone retaining walls, known as spandrels. The roadway was built on top. Stone arch bridges are the oldest bridges still in existence in Maryland. Plentiful native stone made them popular in the north-central and western areas of the state.

Timber Bridges

Maryland's abundant forests provided ample resources for wooden bridges, the first bridge type built in the state. Wooden bridges were reliable, cheap, and easy to build, but they weathered rapidly. The earliest were as elemental as logs lashed together to form simple beam bridges. Later, supports made by hammering boards into triangles allowed timber to span greater distances than a single beam could. Beginning in the 1800s, wooden bridges were covered to protect them from deterioration. They were supported by various truss types: the king-post truss (one large wooden triangle), multiple king-post truss (a series of smaller triangles), queen-post truss (a lengthened version of the king-post truss), or Burr arch truss (a wooden arch combined with a multiple king-post truss).

Concrete Bridges

The use of local materials and local labor made concrete bridges popular in Maryland in the Great Depression of the 1930s. Reinforced concrete was fashioned into beams, slabs, and arches and was particularly suitable in standardizing bridge design. All of Maryland's concrete arches are deck arches, meaning the arch is located below the roadway. If the walls of the arch are solid, it is called a closed-spandrel arch; if the walls are opened into ribs or columns, it is called an open-spandrel arch. Where aesthetics are important, open-spandrel concrete arches are often preferred. Closed-spandrel designs are sometimes faced in stone or sport decorative parapets. A popular closed-spandrel form is the Luten arch, recognized by its very narrow crown and long, shallow arch.

Metal Truss Bridges

Iron and steel truss bridges are among Maryland's most familiar historic bridges. Truss bridges are defined by their arrangement of triangular panels, which work in compression or tension against the forces of gravity. The most widely used in Maryland and the simplest was the Pratt truss, which has diagonal elements in tension and vertical elements in compression. Its numerous variations include the Whipple, or double-intersection Pratt (whose diagonal elements extend over two panels); the Parker and the camelback (both with an arching top chord); the bowstring (shaped like a bow); the Warren (a series of equilateral triangles); and the Howe (the opposite of a Pratt). A through truss rises above the road and has overhead bracing; a pony truss has no overhead bracing; a deck truss is entirely below the road. Early metal trusses were connected with pins and eyebars; later they were riveted and then bolted.

Metal Suspension, Arch, and Cantilever Bridges

Metal suspension, arch, and cantilever bridges are all suitable for long-span crossings such as the Potomac and Susquehanna Rivers and the Chesapeake Bay. In a metal suspension bridge, the roadway is supported from above, rather than below. It hangs from suspenders, or metal rods, attached to cables draped over towers, which are attached to heavy masonry anchorages. A metal arch bridge consists of two or more parallel arches of iron or steel. If the arches are above the roadway, they are through arches. If the arches are below the roadway, they are deck arches. A cantilever bridge is a type of metal truss supported in the middle or at one side, rather than at both ends; the bridge extends, or cantilevers, beyond the supporting pier.

Movable Bridges

Because navigation was so vital in Maryland, especially in the Tidewater region, ship passage was an important consideration for bridge builders. When it was not practical to build high enough for ship clearance, movable bridges—which have one or more spans that open in addition to several fixed approach spans—were constructed. Movable bridges in Maryland are one of two types: bascule bridges, also known as drawbridges, which have one or two leaves or deck sections that lift upward by mechanical means; or swing spans, which rotate horizontally around a central pier. Most of Maryland's movable bridges were built in the eastern part of the state.

The Eastern Shore

Pocomoke City Bridge · 1920, Pocomoke, Worcester County

⬅️ The classical revival Pocomoke City Bridge—a double-leaf trunnion bascule, or drawbridge—was built with an eye to both beauty and utility. Once the primary method of spanning the Eastern Shore's navigable rivers, drawbridges are today considered an impediment to rapid travel and are fast becoming a rarity.

Chester River Bridge · 1930, Chestertown, Kent County

⬆️ Maryland's bridge builders confronted varied geographical challenges. During construction of the Chester River Bridge, a double-leaf bascule, multiple piers were driven into the soft riverbed to distribute the load. Rehabilitated in 1976 and 1990, the bridge has been significantly altered. It carries Maryland Route 213.

The Eastern Shore

Carved from a peninsula shared with Delaware and Virginia, the Eastern Shore is dominated by water: the Atlantic Ocean on the east and the Susquehanna River and the Chesapeake Bay on the west, plus the numerous estuaries that course through the landscape. Shipbuilding, fishing, and canneries prosper along the deeply indented coastline, and movable bridges—drawbridges and swing spans— make certain that the Chesapeake Bay is always within reach.

In early colonial days, travel was easier by water than by road, and the area thrived. But as the population spread inland, the bay became a barrier instead of a bond, and the region withdrew into isolation. That changed dramatically in 1952, when Maryland unveiled the four-mile-long Chesapeake Bay Bridge (renamed for Governor William Preston Lane Jr. in 1967). With the opening of this engineering marvel—the world's largest overwater steel structure at that time—the Eastern Shore's historic and nearly pristine isolation came to an end. The region would never be quite the same.

Dozens of smaller, little-known spans knit together remote pockets of the Eastern Shore. Most of the state's movable bridges are found in this region, where life revolves around the tides. Navigation is essential to move local farm and maritime products to market, so bridges must be high enough for boats to clear them or must be able to open. Because high, fixed bridges required expensive approach work, movable bridges became the primary method of spanning the Eastern Shore's navigable rivers.

On the lower Eastern Shore, parts of which remain rural and isolated, short-span timber-beam bridges take roads over broad marshy creeks and rivers of grass. The upper Eastern Shore, in contrast, is located at the headwaters of the Chesapeake Bay on the crossroads between Baltimore and major eastern cities and is slightly higher in elevation than the rest of the region. Dairy farms and corn, soy, and grain fields dominate the landscape. From the mid- to the late 1800s, covered wooden bridges stood near mill sites, such as the one at Gilpin's Falls. Later, more reliable metal trusses took their place.

Steamboats historically handled most of the transport along the shore, but with the decline of the steamboat and the rise of the automobile in the 1920s and 1930s, trucks became the major carriers of the region's agricultural and maritime produce. Highway transport was faster but required better roads and bridges to handle the increase in traffic. In the 1930s Maryland began a bridge-building program to replace narrow and unsafe bridges on the Eastern Shore's major highways, resulting in several new spans, including the Thomas J. Hatem Memorial Bridge (1939–40) over the Susquehanna River at Havre de Grace.

Gilpin's Falls Covered Bridge · ca. 1860, Bayview, Cecil County

Hurried travelers on Maryland Route 272 zoom past this
simple relic, which sits over Northeast Creek just east of the
busy highway. At 119 feet, Gilpin's Falls Bridge is the longest remaining wooden covered bridge in the state. Built near the site of several
former mills, it is a type known as a Burr arch, the support system
patented in 1817 by the Connecticut bridge builder Theodore Burr,
who added an arch to triangular trusses to make a bridge stronger.
Inside its sheltering clapboard walls, the arched timber beams,
encased in a multiple king-post truss, stretch from bank to bank.
Gilpin's Falls Bridge was restored after a century of use but no
longer handles automobile traffic. Today it is open to pedestrians.

Bell Manor Road Bridge · ca. 1885, Conowingo, Cecil County

The upper Eastern Shore's rolling farms, fields, and sleepy rural towns were a thriving part of the Maryland economy in the nineteenth century. But seasonal floods routinely swept away or severely damaged wooden plank bridges, no matter how sturdy they seemed. In the 1880s progressive Cecil County commissioners launched a campaign to build new, more reliable metal trusses. Several were designed by Charles H. Latrobe, including Bell Manor Road Bridge. Since 1885 this bridge has carried traffic over the slow-moving Conowingo Creek just before it spills into the mighty Susquehanna River. Only a handful of cars each day now make the journey over the 105-foot-long iron Pratt through truss, a favorite of local anglers.

Rolling Mill Road Bridge · ca. 1885–1900, North East, Cecil County

Also believed to be part of the county's truss-building campaign of the 1880s is Rolling Mill Road Bridge, a single-span Pratt through truss about 60 feet long. The old industrial community of North East, where an iron forge and a rolling mill flourished in the mid-nineteenth century, takes its name from the Northeast Creek that passes beneath the bridge. The delicate four-panel truss is probably the work of Charles H. Latrobe, at one time a partner in Smith, Latrobe and Company, one of the first companies organized in Maryland to design, manufacture, and erect metal bridges. When the firm dissolved in the 1880s, Latrobe launched a solo career, working in rural Maryland as well as in Baltimore.

McCauley Road Bridge · ca. 1885–1900, Oakwood, Cecil County

Basin Run, a tributary of Octoraro Creek near the historic community of Rowlandsville, founded around 1749, runs below this single-span Pratt pony truss. Sturdy granite and field-stone houses were built along the creek, and the area bustled with gristmills, sawmills, and iron mills, prompting the county to construct this 53-foot-long bridge about 1885. Just south of town, the dramatic ruins of an elevated train bridge that brought the Penn Central Railroad to the busy town are a further reminder of the passage of time.

Pocomoke City Bridge · 1920, Pocomoke, Worcester County

Maritime enterprises on the Eastern Shore require speedy transport to market. Water is king: highway bridges must make way, and drivers must wait while boats sail through. When the picturesque, whitewashed Pocomoke City Bridge opened in 1920, it was the major entrance to this charming maritime town. With its elegant Doric columns and Classical Revival tender's house, the 275-foot-long drawbridge, part of Maryland Route 675, must have made a pleasing first impression. A double-leaf trunnion bascule, the bridge opened regularly for the classic steamer ships that plied the Pocomoke River in the early twentieth century. Today a boardwalk follows the riverbank on the bridge's east side.

Dover Bridge · 1932, Easton, Talbot County

The 65-foot-long central span of Dover Bridge, which carries Maryland Route 331, swings open for the pleasure boats and barges that chug along the Choptank River out to the Chesapeake Bay. Between 1904 and 1939 the Maryland State Roads Commission built at least seventeen swing spans over navigable waters, including this one, constructed in 1932. Swing spans were preferred by many engineers because they were simple, reliable, and economical. Built by the J. E. Greiner Company, this three-span steel Warren truss is one of only four historic swing spans left in the state. The tender's house of most movable bridges is located on the bridge itself, but here the engineers sited it on the riverbank.

Snow Hill Bridge · 1932, Snow Hill, Worcester County

Time seems to stand still in Snow Hill, which looks and feels like a peaceful nineteenth-century village. Located at the head of the Pocomoke River close to the Atlantic Ocean, it was once an important Eastern Shore commercial center. Historic photographs reveal little change since this single-leaf bascule, one of the simplest and smallest of Maryland's movable spans, was built across the Pocomoke River in 1932. Carrying Maryland Route 12, it measures only 90 feet long, including the approach span. Its neoclassical concrete tender's house is unoccupied; boaters must call in advance to schedule an opening.

Cambridge Bridge · 1939, Cambridge, Dorchester County

Cambridge Bridge opens several times a day for crabbers making their way into the Chesapeake Bay. The 371-foot-long bridge, a double-leaf rolling-lift bascule, was built when the narrow 1870 bridge over Cambridge Creek became overwhelmed with automobile traffic. The end of the steamboat era and the increase in car and truck traffic in the 1930s prompted Maryland to replace many narrow, unsafe Eastern Shore bridges. Cambridge, one of the state's oldest cities, is home to a cottage shipbuilding industry, fruit and vegetable canneries, and oyster packing, all of which benefitted from construction of the bridge, which carries Maryland Route 795.

Bestpitch Ferry Road Bridge · 1946, Hare Town, Dorchester County

Few timber-beam bridges remain in Maryland, most of them having been succeeded by far more durable steel or concrete bridges. However, for the soft muddy soil of the Tidewater region, short timber-beam spans still seem the best choice. A wilderness of marshland surrounds the eleven-span Bestpitch Ferry Road Bridge, which takes Bestpitch Ferry Road over the Transquaking River in the isolated low country of Maryland. Built in 1946, this 192-foot-long timber-beam structure is one of Maryland's youngest historic bridges, yet its location in a sea of marsh grass and weeping willows

Conowingo Dam Bridge · 1927, Conowingo, Harford County

Built by the Philadelphia Electric Power Company in 1927 to generate power for Pennsylvania, the mighty Conowingo Dam carries U.S. Route 1 across the Susquehanna River about ten miles upstream from the Chesapeake Bay. Reputedly the longest concrete-slab dam in the country—it measures 114 feet high and approximately 4,700 feet long—the Conowingo Dam backs up the river for more than fourteen miles to form Conowingo Lake. Fifty-three concrete-beam spans form the bridge atop the dam. The structure's Art Deco styling is the work of the Boston-based firm Stone and Webster.

Thomas J. Hatem Memorial Bridge · 1939–40
Havre de Grace, Harford County

The point where the Susquehanna River meets the Chesapeake Bay has been a strategic crossing point for generations. Ferries made the long trip across until 1910, when local citizens converted a railway bridge into a toll bridge. It proved to be a gold mine but inadequate for the traffic, even with the addition of a second level. In 1940 the state erected this high, fixed steel truss to make the trip on U.S. Route 40 safer and easier. Designed by the J. E. Greiner Company, the combination through-and-deck arch extends 7,618 feet and measures 177 feet from the water to its highest point. In 1986 it was dedicated to Hatem, a distinguished Harford County citizen.

Chesapeake City Bridge
1948, Chesapeake City, Cecil County

In an ever-changing procession, ships, tugboats, and pleasure craft glide beneath the majestic Chesapeake City Bridge, a steel tied-arch bridge that carries Maryland Route 213 high above the Chesapeake and Delaware Canal. In 1942 a tanker bound for Baltimore demolished an earlier lift bridge here. Because of wartime rationing, residents made do with ferry service until the new bridge was finished in 1948. The U.S. Army Corps of Engineers built and maintains the 540-foot-long bridge and monitors traffic through the canal. The bridge rises 135 feet above the water, allowing large ships to pass beneath on this route between Baltimore and Philadelphia.

**William Preston Lane Jr. Memorial Bridge
(Chesapeake Bay Bridge)** · 1949–52 and 1969–73
Anne Arundel and Queen Anne's Counties

A triumph of human ingenuity, the William Preston Lane Jr. Memorial Bridge (U.S. Routes 50/301) spans one of Maryland's great natural barriers, the Chesapeake Bay. Stretching 4.03 miles, this magnificent, curving ribbon of steel is graceful and elegant yet strong enough to carry 24 million vehicles a year. The original bridge (right in the photograph) opened in 1952. Supported by 354-foot-high suspension towers and 14-inch-thick cables, its roadway soars 198 feet above the water, providing ample clearance for ocean-going vessels making their way into Baltimore's port. Designed by the J. E. Greiner Company, the bridge features 123 steel spans—cantilever trusses, simple trusses, plate girders, and beams—culminating in the bridge's centerpiece: a 1,600-foot-long suspension span. In 1967 the bridge was renamed for Lane, governor of Maryland from 1947 to 1951, who vowed to give the state "a system of highways second to none in the nation." To accommodate increased traffic, a parallel bridge, also designed by the J. E. Greiner Company, opened in 1973. The older bridge carries eastbound traffic; the newer one, with its 1,500-foot-long suspension span and 379-foot-high towers, takes traffic west.

Central and Southern Maryland

The image shows a bridge with a sign reading "WALK YOUR HORSES OVER THIS DRAW / UNDER A PENALTY OF 20 DOLLARS FINE"

Howard Street Bridge · 1938, Baltimore City

Baltimore is home to two notable metal arch bridges, including this through arch at Howard Street. The arch has been an enduring bridge form since the ancient Romans built stone arch bridges centuries ago. As technology progressed, the arch was retained, but the materials evolved to include metal and concrete.

Light Street Bridge · 1856, 1891, Baltimore City

One of Maryland's first swing-span movable bridges, the Light Street Bridge, no longer extant, took horses and cable cars across the Patapsco River's Middle Branch. In 1891 the wooden swing span was replaced with an iron one made by the King Manufacturing Company of Cleveland. Timber piers supported the fixed spans.

Central and Southern Maryland

In central and southern Maryland, bridge engineers dreamed big and built boldly, leaving a stunning array of nationally recognized engineering landmarks. Elegant masonry arches, metal trusses, metal arches, and one of the state's rare cantilever spans stretch across the region's rivers and streams that drain into the Chesapeake Bay.

Maryland's largest city, Baltimore, was the focus of the region's bridge building. The destination of several major turnpikes, birthplace of the nation's first railroad, hub of metal truss design, and a major port, it became a laboratory for early American bridge design. Much of the experimentation took place under the direction of the railroad's pioneering engineers, such as Benjamin Henry Latrobe Jr., Wendel Bollman, and John E. Greiner.

The Baltimore and Ohio Railroad built several imposing masonry arch bridges in the region, including two celebrated National Historic Landmark spans still in use today, the Carrollton and Thomas Viaducts. But it was the B&O's experiments with iron that proved most far reaching. The Bollman truss, the first all-iron bridge type to be used extensively by the railroad, represented the key moment when bridge design evolved from rule-of-thumb intuition to scientific engineering. The world's sole remaining Bollman truss bridge stands in Savage, Maryland.

Several national metal truss bridge-building companies originated in Baltimore, including the Bollman Company, which worked in North Carolina, Cuba, and Mexico, and Smith, Latrobe and Company, which built bridges across the Mississippi, Missouri, and Kentucky Rivers. Baltimore emerged as the center of metal truss building in the state. From there the use of metal trusses spread out to other regions as they gained acceptance for roadway use.

When cars and trucks began to edge out horses and buggies and ultimately the railroad early in the twentieth century, reinforced concrete emerged as a major challenger to metal in central Maryland. Strong, with great decorative capabilities, concrete was suitable for standardized designs that met the growing traffic challenges. Beams and slabs were utilitarian favorites, but concrete arches were chosen for prominent locations, such as the Hanover Street Bridge (now the Vietnam Veterans Memorial Bridge) in Baltimore and Sligo Creek Bridge near Takoma Park.

In 1940 another stunning engineering milestone was achieved with the construction of a mighty cantilever bridge over the Potomac River, the Governor Harry W. Nice Memorial Bridge. The new bridge increased southern Maryland's accessibility as well as its population. Despite the growth, the region remains true to its fishing and agricultural roots.

Thomas Viaduct · 1832–35, Elkridge, Howard County

People scoffed when Benjamin Henry Latrobe Jr. began building his 612-foot-long stone bridge over the Patapsco River. "Latrobe's Folly," as some doubters called it, opened in 1835 and has been in continuous service between Baltimore and Washington, D.C., ever since. One of the world's oldest stone multiple-arch railroad bridges, the eight-span National Historic Landmark was named for Phillip E. Thomas, first president of the Baltimore and Ohio Railroad. Standing 59 feet above the riverbed on a four-degree curve, the bridge has carried every type of locomotive from the original six-ton engines of the 1800s to the three-hundred-ton engines of today without requiring any major repairs or modifications, confirming the stunning misjudgment of early skeptics.

Cabin John Aqueduct · 1853–63, Glen Echo, Montgomery County

The Cabin John Aqueduct was the longest single-span stone arch in the world when it was completed, a designation it held for nearly forty years. Designed by the famed engineer Montgomery C. Meigs (1816–92), the 220-foot-long sandstone-and-granite arch carried a roadway and a conduit, which channeled drinking water into Washington, D.C. A lack of funds dragged out the project for ten years. Known originally as Union Arch, it was heavily guarded during the Civil War to protect the capital's water supply. In 1862 feelings against Jefferson Davis, secretary of war during the early part of the bridge's construction, ran so high that his name was chiseled off a plaque; it was restored by President Theodore Roosevelt in 1909. Still in use today, the bridge takes MacArthur Boulevard over Cabin John Creek and Parkway.

Bollman Truss Suspension Bridge · 1869, Savage, Howard County

This is the world's only surviving example of the distinctive iron truss system pioneered by the engineer Wendel Bollman for the Baltimore and Ohio Railroad in 1850. An unusual blend of cast-iron and wrought-iron bracing gives the two-span, 160-foot-long bridge a spider-web look. Bollman's revolutionary design, a combination truss-and-suspension bridge, signaled the widespread use of iron trusses for railroad and later roadway use. This example was constructed on the B&O's main line but was moved to its present location over the Little Patuxent River in 1887, when the railroad opened a spur line to Savage to service a busy cotton mill. Restored in 1968 and now used as a pedestrian bridge, it is a National Historic Landmark.

Hanover Street Bridge · 1916, Baltimore City

At the time of its completion, the Hanover Street Bridge, now known as the Vietnam Veterans Memorial Bridge, was the largest reinforced-concrete bridge in the state. Its thirty-eight spans reach 2,290 feet across the Middle Branch of the Patapsco River in Baltimore's Outer Harbor. Designed by the former Baltimore and Ohio engineer John E. Greiner for the Maryland State Roads Commission, the Beaux Arts–style bridge vastly improved transportation in Baltimore at a time when trucks began to take precedence over the steamboat in getting goods to market. The four concrete towers, positioned at the corners of the steel draw span, give the bridge a classical symmetry.

Guilford Avenue Bridge · 1936, Baltimore City

The City Beautiful Movement greatly influenced bridge design in Baltimore, which in the late 1800s boasted several beautifully ornamented iron arch bridges. Their form is echoed in newer, more muscular-looking steel arches built in the 1930s on Guilford Avenue and Howard Street, the only two metal arch bridges remaining in the city and two of only a handful of such bridges built in the state. The 408-foot-long Guilford Avenue Bridge over Jones Falls replaced an elegant but unstable 1879 cast-iron arch that could not support the heavy trucks that serviced the city's thriving manufacturing district. In 1962 the bridge was shortened from a double-arch to a single-arch bridge when Interstate 83 was routed beneath it.

Howard Street Bridge · 1938, Baltimore City

Between 1880 and 1900 Baltimore constructed a number of important iron arch bridges spanning Jones Falls that were designed by Charles H. Latrobe. As traffic loads increased, they were replaced with bridges such as this steel arch on Howard Street and the similar one on Guilford Avenue, both meant to complement Latrobe's work. The 979-foot-long Howard Street Bridge—which crosses Interstate 83, Amtrak rail lines, and Jones Falls—consists of a double-span through arch and beam spans. Within the arches, large steel suspenders support the roadway, which carries five lanes of traffic into the city's busy commercial center. The protective metal starbursts prevent adventurers from climbing up the arches.

Sligo Creek Bridge

1932, Takoma Park, Montgomery County

This triple-span, open-spandrel concrete arch reaches 225 feet across a deep ravine carved out by Sligo Creek in a woodland near Takoma Park just north of Washington, D.C. The lacy, open-spandrel design lightens the bridge physically and visually, an effect enhanced by the decorative open parapets. This graceful bridge carrying Maryland Route 195 was built as part of an effort by the Maryland State Roads Commission to unify its bridges. In 1932 the state launched a program to build bridges on its secondary and feeder roads, and during the next two years about 170 bridges were built, including this one.

Brighton Dam Road Bridge

1941–44, Brighton, Montgomery County

In the 1940s the burgeoning federal government and war-related industries created a housing boom near Washington, D.C.—and a need to increase the water supply. Brighton Dam held back the Patuxent River to form the Triadelphia Reservoir, which fed the Maryland suburbs. Atop the dam a thirty-span, 600-foot-long concrete-slab bridge links Montgomery and Howard Counties. Beneath the reservoir, also used for recreation purposes, lie remnants of a tiny nineteenth-century mill town founded by three men who married three sisters. They called the town Triadelphia (city of three).

Governor Harry W. Nice Memorial Bridge

1939–49, Newburg, Charles County

Before the opening of this 1.7-mile-long toll bridge in 1949, only ferry service connected southern Maryland to Virginia at this wide crossing of the Potomac River. This bridge did for southern Maryland what the Chesapeake Bay Bridge did for the Eastern Shore: it ended the region's long-time geographical isolation. The majestic structure, which carries U.S. Route 301, is the state's only metal cantilever bridge. Cantilever bridges, in which the roadway cantilevers or extends beyond the piers, were economical for long crossings such as this. In addition to the cantilever span, the bridge includes steel beams, trusses, and plate-girder spans. At the center the roadway soars 135 feet above the main ship channel, allowing plenty of space for navigation. President Franklin D. Roosevelt attended the groundbreaking ceremony.

Northern Maryland

Utica Mills Bridge · 1860, Thurmont, Frederick County

The earliest known covered bridges in Maryland were extremely long multispan structures built over the Susquehanna River in the early 1800s. Most were lost to poor maintenance, fires, or floods. The Utica Mills Bridge over Fishing Creek is more typical of the smaller covered bridges in pastoral settings that survived.

Parkton Bridge · 1929, Parkton, Baltimore County

Bridge building in the late 1920s was labor intensive but offered an honest day's wage when work was hard to come by. These men are building a single-barrel, closed-spandrel concrete arch on Maryland Route 45, a rerouting of the old York Road, which crossed Maryland's oldest extant bridge, the 1809 Parkton stone arch.

Northern Maryland

In the lush green valleys and rolling hills of northern Maryland, the early culture and economy hinged on agriculture and milling. Abundant rivers ripe for turning the mills' giant waterwheels plunge south and east through the region before emptying into the Chesapeake Bay. Bridge building here followed a logical sequence: a natural ford, a mill site, a bridge. A remarkable collection of covered bridges and metal trusses crown these rivers.

The region is home to the most covered bridges in the state. In their heyday, between 1800 and 1900, as many as fifty-two covered wooden bridges graced the Maryland landscape. Today only a handful remain. Three covered bridges—Roddy Road, Utica Mills, Loys Station—are located in Frederick County; a fourth, Jericho Road Bridge, links Baltimore and Harford Counties. Hand built by local craftsmen, they provide an excellent comparison of the various truss types used by nineteenth-century bridge makers: the king-post truss, a large triangle with one span; the multiple king-post truss, a series of smaller triangles within one span; and the Burr arch, a graceful support system in which an arch is encased in a multiple king-post truss.

As iron became available after the Civil War, local communities began the drive to replace wooden bridges with metal trusses. Metal bridges were more durable, cheaper, and more expedient than carpenter-built bridges. They were also largely immune to the dangers of floods, fires, and other disasters that befell wooden bridges. By assuring travelers of safe and dependable passage throughout the year, metal trusses were important in the development of this pastoral region. An exceptional group of metal trusses is found in Frederick County.

Once a community decided that it needed a metal bridge, county commissioners advertised for bids in the local newspaper. Bridge manufacturers from New York, Ohio, Pennsylvania, and Maryland selected a suitable design from among their patented list and submitted it. After the bid was awarded, the bridge was manufactured and then shipped to the site to be assembled by local crews. The Wrought Iron Bridge Company of Canton, Ohio, was especially active in this area.

A modern iron truss became the pride of every small town. But early in the twentieth century, standardized concrete bridges became more popular for small to moderate crossings. The metal truss still remained the bridge of choice for large crossings; an example is the sturdy Point of Rocks Bridge over the Potomac River. But the new trusses were made of steel and were heavier and more solid than their iron predecessors.

Roddy Road Covered Bridge

ca. 1850, Thurmont, Frederick County

Covered bridges may be the best-loved historic American bridge type. Three of Maryland's remaining covered bridges stand within a fifteen-mile radius in Frederick County, while a fourth connects Baltimore and Harford Counties. All cross small streams and fit perfectly into this rolling countryside of fields and farms (although one is now sited in a park). Roddy Road Bridge, just 40 feet long, is the smallest of Maryland's covered bridges. Clad in red beveled boards and topped by a tin gabled roof, the one-lane bridge is supported by a single king-post truss. Owens Creek flows below.

Utica Mills Covered Bridge

1850, Thurmont, Frederick County

Thrifty Maryland forebears saved this bridge from complete destruction when a flood swept half of it away in 1889. The bridge originally spanned the Monocacy River on Devilbliss Road, but after the flood the remainder was moved by wagon to Utica, about two and a half miles away, and reassembled over Fishing Creek. Constructed of large, hand-hewn timbers, this 100-foot-long bridge is supported by a Burr arch truss. Patented by the inventor Theodore Burr, this truss combines a series of triangles with a long wooden arch to make the bridge more rigid and thus stronger.

Jericho Road Covered Bridge
1865, Kingsville, Baltimore County

Well over a hundred years ago, frugal citizens in the historic villages of Jericho and Jerusalem petitioned Maryland's General Assembly for a new bridge over Little Gunpowder Falls that would connect their towns while bypassing the toll roads and turnpikes nearby. On completion in 1865, Jericho Road Bridge helped this center of cotton mills, flour mills, and ironworks flourish. Sheathed in cedar boards and supported by a timber-beam Burr arch, the single-span 100-foot-long bridge links Harford and Baltimore Counties. It was restored in 1982 and is still in use carrying Jericho Road over the stream.

Loys Station Covered Bridge
1900, Thurmont, Frederick County

Taller and narrower than Maryland's other covered bridges, Loys Station crosses Owens Creek at the north edge of a peaceful, tree-filled park. The 90-foot-long twin span was named for a nearby train stop. Quite a few cars rumble through the one-lane structure on Old Frederick Road, but the bridge's sturdy multiple king-post truss handles the load. The earliest timber bridges in Maryland did not have covers, but bridge builders soon realized that the addition of a protective cladding helped the bridge last far longer. Although designed to be purely practical, the cladding gives the bridge a sense of mystery.

Parkton Stone Arch Bridge · 1809, Parkton, Baltimore County

Maryland's oldest surviving bridge, the Parkton Stone Arch Bridge is a rare relic of the turnpikes that provided the state's first reliable overland transportation. Nearly two centuries old, the classic two-arch stone bridge was one of five built on the Baltimore and York Turnpike between 1800 and 1810. Spanning Little Gunpowder Falls, the 37-foot-long bridge was probably designed by the British-born engineer John Davis (1770–1864). In 1910 the State Roads Commission purchased the turnpike and began to make improvements, but the growing number and size of vehicles made each upgrade obsolete. By the 1930s the road and bridge were bypassed altogether. The bridge now sits on a dead-end village street.

LeGore Bridge · 1900, Woodsboro, Frederick County

James W. LeGore could hardly have envisioned the outcome of his decision to build this monumental stone bridge over the Monocacy River. Hoping to expedite deliveries to his nearby quarry, he put up $100,000 of his own money and used his own men to build the 248-foot-long engineering marvel. Four- and six-horse teams hauled limestone from the quarry to the river, where laborers spent four years creating the bridge's five deep arches and massive piers. The county initially promised to repay LeGore but later reneged. During construction, malcontents tried unsuccessfully to dynamite one of its main piers, and later LeGore's son committed suicide by jumping from the bridge. Part of LeGore Bridge Road, it was restored in 1981.

Detour Road Bridge · ca. 1872, Catoctin, Frederick County

For more than one hundred years this 92-foot-long iron bow-string pony truss carried horses, wagons, and Model Ts across Big Pipe Creek near the town of Detour in Carroll County. Patented by the inventor Squire Whipple (1804–88) in 1841, the bowstring truss—its semicircular shape is similar to that of a bow—was inexpensive, lightweight, sturdy, and perfect for rural crossings. This bridge was manufactured by the Wrought Iron Bridge Company of Canton, Ohio, which offered many types of trusses. In 1977, when the one-lane bridge was replaced by a three-span steel-beam bridge, its arches were loaded onto a trailer and moved with police escort to a new home over Little Hunting Creek. Reduced in width from 13 to 8 feet, the bridge was given new life as a pedestrian crossing in Cunningham Falls State Park.

Four Points Bridge · ca. 1876, Emmitsburg, Frederick County

This iron span, a half-hip Pratt through truss, takes the Four Points–Keysville Road over Toms Creek near Emmitsburg. A subtype of the popular Pratt truss, the half-hip conserved material by eliminating some vertical members. The 103-foot-long bridge is one of many in the state designed and built around 1876 by the prolific Wrought Iron Bridge Company of Canton, Ohio, one of the largest manufacturers of iron truss bridges in the nineteenth century. In its heyday the company employed 270 men and had projects in twenty-five states. Four Points Bridge's ornate name plates and decorative portals are reminiscent of those of the Poffenberger Road Bridge, which may have been built by the same company.

Poffenberger Road Bridge · 1878, Burkittsville, Frederick County

The Poffenberger Road Bridge may be Maryland's only surviving example of a Whipple truss, also known as a double-intersection Pratt because extra diagonal members intersect two panels. Circular decorations of thin sheet metal ornament the portals of this one-span, 120-foot-long bridge, which crosses Catoctin Creek. Designed by either the Wrought Iron Bridge Company or the Penn Bridge Company, this bridge was probably built in part to service the nearby Lewis Mill.

Bennies Hill Road Bridge · ca. 1880, Middletown, Frederick County

On a narrow winding path, surrounded by woods and pasture, Bennies Hill Road makes a sharp turn onto the wooden deck of a picturesque bowstring pony truss. Just a single lane wide and 94 feet long, it delivers travelers across Catoctin Creek. The bowstring arch was a popular design for the King Iron Bridge Manufacturing Company of Cleveland, which designed and built this bridge around 1880. In 1854, under the leadership of founder Zenas King, the company was building about twenty-five bridges a year; by 1874 it was producing between 250 and 300 and was on its way to becoming one of the country's leading bridge manufacturers. Few bowstring trusses were built in Maryland. This is the only one still in its original location.

Noble's Mill Bridge · 1883

Darlington, Harford County

With the addition of a horse-drawn carriage, this 1883 metal truss bridge and adjacent stone mill could be a scene from Currier and Ives. Named for Benjamin Noble, whose gristmill processed forty barrels of flour a day with power from the swift-flowing Deer Creek, this single-span Pratt through truss helped the mill remain competitive. Built by the Wrought Iron Bridge Company of Canton, Ohio, the 148-foot-long bridge is part of Noble's Mill Road, which leads past the mill.

Simpson's Mill Road Bridge · ca. 1890–1900

Johnsville, Frederick County

A remote location deep in farmland has helped preserve Simpson's Mill Road Bridge as well as many other small nineteenth-century trusses throughout Maryland. A 90-foot-long single-span Pratt through truss, it carries Simpson's Mill Road over Little Pipe Creek. Pratt trusses were extremely popular because of their strength, economy, and ease of assembly. Patented by Caleb and Thomas Pratt (1812–75) in 1844, the Pratt is Maryland's most common surviving truss type.

Bullfrog Road Bridge
1908, Emmitsburg, Frederick County

The steel Bullfrog Road Bridge is the only Parker through truss in Frederick County and may be the oldest one in the state. A variation of the more common Pratt truss, the Parker, patented by C. H. Parker, has a polygonal top chord that gives the bridge a rounded appearance. The Parker is similar to a camelback truss, but the angle of the top chord changes at every panel. Because of its strength, the Parker was a popular choice for longer spans, such as this 179-foot-long crossing over the Monocacy River. The builder, York Bridge Company of York, Pennsylvania, shipped at least twenty bridges to Frederick County in the early 1900s.

Pearre Road Bridge
1908, New Windsor, Carroll County

In another storybook setting—deep in the woods near a nine-teenth-century stone mill—a one-lane steel pony truss carries Pearre Road over Sams Creek, a natural boundary that divides Car-roll and Frederick Counties. This 1908 bridge, just 63 feet long, is one of Maryland's rare Warren pony trusses. Patented by the British engineer James Warren and Theobald Monzani in 1846, the Warren truss—formed by equilateral triangles—was not widely used until the early twentieth century. Only a handful of Warren trusses remain in Maryland. The Pearre Road Bridge was also built by the York Bridge Company of York, Pennsylvania.

Warren Road Bridge · 1922, Cockeysville, Baltimore County

This lacy steel truss with elaborate portals, designed at a time when labor was cheap and decorative elements flourished, carries Warren Road over Loch Raven Reservoir, a source of Baltimore's municipal water supply since the 1880s. Enlarging the reservoir to meet Baltimore's increasing water needs in 1922 required four new roads, four new bridges, the relocation of five miles of railroad tracks, and the removal of two villages. Designed by the Phoenix Bridge Company of Phoenixville, Pennsylvania, this three-span Parker through truss measures 624 feet long. Loch Raven Reservoir also supports a recreational area offering fishing, boating, hiking, and picnicking.

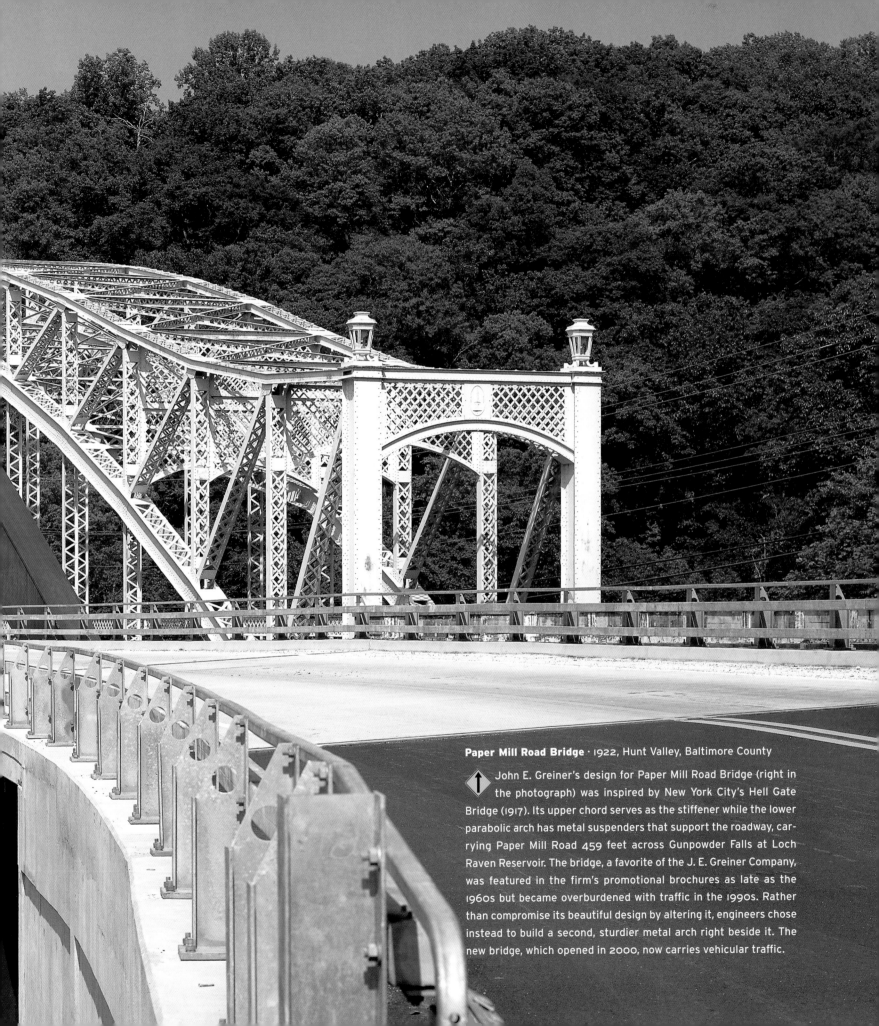

Paper Mill Road Bridge · 1922, Hunt Valley, Baltimore County

John E. Greiner's design for Paper Mill Road Bridge (right in the photograph) was inspired by New York City's Hell Gate Bridge (1917). Its upper chord serves as the stiffener while the lower parabolic arch has metal suspenders that support the roadway, carrying Paper Mill Road 459 feet across Gunpowder Falls at Loch Raven Reservoir. The bridge, a favorite of the J. E. Greiner Company, was featured in the firm's promotional brochures as late as the 1960s but became overburdened with traffic in the 1990s. Rather than compromise its beautiful design by altering it, engineers chose instead to build a second, sturdier metal arch right beside it. The new bridge, which opened in 2000, now carries vehicular traffic.

Point of Rocks Bridge · 1939, Point of Rocks, Frederick County

Like other strategic fording locations, Point of Rocks along the Potomac River was the site of military maneuvers during the Civil War. Today an impressive eight-span steel bridge, each span 165 feet long, carries U.S. Route 15 over this major crossing. Built in 1939, this bridge is a type known as a camelback truss, so called because the upper chord of each span is polygonal, consisting of five slopes, and the resulting arch resembles a camel's back. The bridge's sturdy construction reflects the evolution of truss design in response to the heavier load requirements of the automobile. At this point the river collects a great deal of debris, which is deflected by the bridge's substantial concrete piers.

Pretty Boy Dam Bridge · 1932, Pretty Boy, Baltimore County

Dams can be both utilitarian and picturesque, as Pretty Boy Dam makes clear. It impounds Gunpowder Falls into a large reservoir that supplies the city of Baltimore with fresh water. Four concrete arches and twenty-six concrete girders reach nearly 700 feet over the mouth of the reservoir, permitting traffic on Pretty Boy Dam Road to cross. Decorative parapets with Classical Revival details add to the aesthetics. A gatehouse at midspan provides access to the dam. Funds to build the dam came from a public improvement commission, which for a nine-year period beginning in 1920 raised money from loans approved by Baltimore citizens.

Glyndon Bridge · 1947, Glyndon, Baltimore County

Noted for its stylish design, Glyndon Bridge is a rare example of a Moderne concrete-slab bridge. The essence of the style is the expression of forward motion: the bridge's aerodynamic parapets flare slightly away from the road and then curve back like a scroll to enclose four large concrete urns. The elegance of the bridge's appearance may be due to the fact that the design team included an architecture firm as well as an engineer. Sheathed in a coursed rubble stone veneer, the 214-foot-long bridge carries Maryland Route 128 in five spans over the Western Maryland Railroad rail lines. The arrival of the railroad in 1860 and later the automobile shaped Glyndon's growth from a summer resort to a year-round community.

Western Maryland

U.S. Route 40 over Conococheague Creek and Wilson Bridge

1936 and 1817–19, Wilson, Washington County

An open-spandrel concrete arch majestically frames the nearly two-hundred-year-old Wilson Bridge, which was supplanted in 1936 as the major crossing over Conococheague Creek. The pair offers a glimpse of the evolution of bridge design and materials.

Harpers Ferry Railroad Bridge

1862, Sandy Hook, Washington County

Blown up during the Civil War and often damaged by floods, this Baltimore and Ohio Railroad bridge over the Potomac River, a multispan Bollman truss, was rebuilt many times. Converted to vehicular use in 1936, it no longer stands.

Western Maryland

The rugged landscape of Maryland's western reaches is veined with narrow post roads, sleepy picturesque towns, and forest-clad mountains. For centuries these mountains and the tumbling streams that flow through them proved a thorny obstacle for travelers and bridge makers. Yet the broad, limestone-ridged valleys yielded the perfect material for making strong bridges.

From the earliest days, Marylanders and other settlers sought a way through the mountains into the frontier markets of the Ohio River valley. In 1806 the area became the official gateway to the West with the authorization by Congress of the nation's first federally financed road, which began in Cumberland, Maryland, and shot west over the Alleghenies to Vandalia, Illinois. Beginning in 1811, teams of men with picks and shovels, oxen and horses cleared the roadway, climbing and descending one mountain after another until the entire National Road was finished in 1818.

The new National Road (or Cumberland Road, as Congress always referred to it) and a growing network of semiprivate turnpikes triggered a major bridge-building effort in the region. The resulting stone arch bridges from the nineteenth century, one of the finest collections in the country, became a vital link in the nation's western expansion. Built on natural fords and sites of local commerce, they were regarded as the wonders of their day.

The first, the Casselman River Bridge, was the largest stone arch in the nation when it was built in 1813 and remains a splendid example of early American engineering prowess. The rest of the region's approximately thirty stone arch bridges that still stand were constructed between 1819 and 1863, many at mill sites, and crossed two major creeks, the Antietam and the Conococheague, and their tributaries. The bridges' rustic stonework recalls the Revolutionary War–era farmhouses and stone barns that still dot the limestone-covered hills. Although similar, no two are exactly alike. Parapets are peaked, rounded, undulating, or flat; piers are rounded, flattened, or pointed (to divide the current and deflect logjams and ice floes). All four bridges over the wide, slow Conococheague have four or five arches, while the narrow, deep Antietam has no five-arch spans and only one four-arch bridge. Those across smaller tributaries have one, two, or three arches. Most of Maryland's stone arch bridges have been rebuilt to some extent at some time.

These survivors from the horse-and-buggy era are a testament to the skills of their makers. They are reminders of the era of the National Road and America's relentless push for new opportunity. Two newer bridges in the region, an open-spandrel concrete arch built on U.S. Route 40 in 1936 and the steel tied-arch Blue Bridge in Cumberland, completed in 1954, depict the progress in materials and design that engineers have made since the state's earliest settlers began to demand reliable river crossings.

Casselman River Bridge · 1813, Grantsville, Garrett County

One of the first major bridges on the National Road, the Casselman River Bridge, at 354 feet, was the longest single-span stone arch in the United States when built. Skeptics were sure it would collapse when the supports were removed, but the 80-foot-long arch proved equal to the tidal wave of stagecoaches and cargo wagons that poured over it, carrying goods and people from Cumberland to the western frontier. Today the National Historic Landmark bridge is a pedestrian crossing in a state park. From the bridge one can see the march of progress: a 1930s metal truss on U.S. Route 40, which supplanted the 1813 stone arch, and a modern steel-beam bridge, which carries Interstate 68 and the bulk of traffic in far western Maryland.

Wilson Bridge · 1817–19, Wilson, Washington County

Wilson Bridge, the oldest and perhaps the most graceful of Washington County's limestone arches, was considered a modern marvel at the time of its completion in 1819. Its fine workmanship, by the mason Silas Harry, served as a pattern for some thirty stone bridges that would follow in the region. The five-arch, 210-foot-long bridge carried the state-chartered National Pike (later designated U.S. Route 40) over Conococheague Creek to join the National Road at Cumberland, providing a pivotal link between eastern seaport cities and western markets. Wilson Bridge was bypassed in 1936, when U.S. Route 40 was rerouted and a newer bridge built downstream. It remained open for local traffic until 1972, when flood damage from Hurricane Agnes relegated it to foot traffic.

Roxbury Mill Bridge · 1824, Funkstown, Washington County

Finely crafted of local limestone, this three-arch bridge takes Roxbury Road over Antietam Creek south of Funkstown. Situated near the Roxbury sawmill, a site later occupied by a large distillery complex, it measures 169 feet long. Piers on both sides of the bridge have a rounded or conical shape, although other limestone bridges in the region have pyramidal-shaped piers. Sources vary as to the bridge's designer: it was either the mason Silas Harry or James Lloyd, of the Lloyd bridge-building firm of Pennsylvania. The ruins of two stone structures associated with the distillery complex are located along the west bank of Antietam Creek.

Broadfording Road Bridge · 1829, Cearfoss, Washington County

As Conococheague Creek winds southward toward the Potomac River, it slows in a wide shallow turn that has provided a natural "broad fording" since colonial times. Broadfording Road, in existence since 1747, was one of the region's earliest thoroughfares. The crossing became infinitely easier and more reliable in 1829, when this sturdy 220-foot-long stone bridge was erected. Workers floated local limestone downstream to be fashioned into four arches, two large and two small. This location was the site of a small nineteenth-century community and mill.

Hitt Bridge · 1830, Keedysville, Washington County

Three stone arch bridges over Antietam Creek played crucial roles in the famous Civil War battle of 1862: Burnside Bridge, also known as Lower Bridge; the now-demolished Middle Bridge; and Upper Bridge, known as Hitt Bridge after a nearby mill and also as Hicks Bridge. Union troops camped near Hitt Bridge on the eve of the battle. During and after the conflict, a house and a barn nearby became a makeshift Union hospital. The 105-foot-long bridge, a three-arch span, was built by Silas Harry, an agent for John Weaver, and carries Keedysville Road.

Antietam Iron Works Bridge · 1832, Antietam, Washington County

The only four-arch stone span over Antietam Creek, this 160-foot-long bridge was built by John Weaver at the site of a bustling ironworks complex that supplied cannon to the Continental Army during the Revolutionary War. Two iron furnaces, several lime kilns, a nail factory, a sawmill, and a gristmill operated here intermittently from 1765 through 1886, at their peak employing hundreds of workers and turning out four hundred kegs of nails daily. The bridge takes Harpers Ferry Road over Antietam Creek about one-fourth mile east of the Potomac River.

Booth's Mill Bridge · 1833, Boonsboro, Washington County

Families picnic and children romp along the grassy banks of Devil's Backbone County Park, where Booth's Mill Bridge spans a wide, languorous section of Antietam Creek. This three-arch, 120-foot-long stone bridge, built near Booth's Mill by Charles Wilson in 1833, replaced a wooden span. The stone bridge, part of Maryland Route 68, is also known as Delemere Bridge after the nearby Delemere mansion, which the Reverend Bartholomew Booth operated as a boys' school in the 1770s; one of his more notable pupils was Benedict Arnold's son. The bridge was rehabilitated in 1996.

Pry's Mill Bridge · after 1835, Keedysville, Washington County

This narrow two-arch, 127-foot-long bridge stretching over Little Antietam Creek was built of local limestone by John Burgan. As with most stone arch bridges, the single pier on the upstream side is pointed like the prow of a ship to deflect logjams and ice floes; on the downstream side, it is rounded. Near the bridge, part of Keedysville Road on the outskirts of Keedysville, both a gristmill and a sawmill once operated. Many of the region's major roadways led to and from mill sites, which served as important centers for trade and social meetings. A new bridge played a key role in promoting a community's growth and development.

Burnside Bridge

1836, Antietam National Battlefield, Washington County

In 1836 the master bridge builder John Weaver put the finishing touches on Lower Bridge, a beautifully proportioned, 192-foot-long stone bridge over Antietam Creek. Twenty-six years later the bridge became the battleground for the bloodiest day of fighting in the Civil War. On that pivotal day, September 17, 1862, nearly five thousand Americans lost their lives in the rolling hills and farmland near Sharpsburg. Ever since, the bridge has borne the name of General Ambrose P. Burnside, commander of the Union troops that stormed the bridge under withering Confederate fire. Now under the care of the National Park Service, the three-arch bridge has been faithfully restored to its original condition, including the wooden coping that tops its walls.

Rose's Mill Bridge

1839, Funkstown, Washington County

 A historic plaque affixed to Rose's Mill Bridge, a three-arch, 132-foot-long bridge built to service a grain mill on Antietam Creek, credits John A. Wever as the builder. Historians are not certain if he is the same John Weaver who built many other Washington County stone arch bridges. The widened platform at the southwest corner permitted wagons to load and unload from the building's second-story door. The westernmost arch extends over dry land today but originally spanned the adjacent millrace. Crumbling remnants of a few stone walls are poignant reminders of the thriving mill that brought wagons clattering over the bridge, part of Garis Shop Road.

Claggett's Mill Bridge and Mill Race Bridge

1840 and ca. 1841, Funkstown, Washington County

⬆ Two adjoining stone bridges take Poffenberger Road over Antietam Creek: Claggett's Mill Bridge (right in the photograph), a 173-foot-long, three-arch span with pointed piers erected by John Weaver in 1840, and a 53-foot-long single arch built over the millrace about a year later. The bridges' exquisite stonework mirrors that of Valencia, the Claggett family's eighteenth-century estate northwest of the bridge. The county commissioned the main bridge, but the Claggetts probably paid for the smaller section over the millrace. Weaver may have built both, but the keystone in the millrace arch suggests that it is the work of a Chesapeake and Ohio Canal mason.

Felfoot Bridge

ca. 1850, Keedysville, Washington County

⬦ Echoing the stonework of its eighteenth-century namesake, Felfoot Farm, this 90-foot-long, two-arch bridge, built by George Burgan, carries Dogstreet Road over an established fording spot on Little Antietam Creek between two historic mill sites: Nichodemus Mill to the east and Hess Mill to the west. At the time Dogstreet Road was a significant thoroughfare in southern Washington County, connecting Middletown to Sharpsburg. The squared pilasters, which project out from the bridge, look like supporting columns but are decorative rather than structural. The Felfoot Farm complex, east of the bridge, includes a stone-and-brick house, a stone barn, and outbuildings.

Old Forge Road Bridge

1863, Hagerstown, Washington County

⬆ Named for an eighteenth-century forge operated by the Hughes family, this three-arch span on Old Forge Road over Antietam Creek, erected by W. H. Eirley, is the youngest of Washington County's dated stone bridges. The eastern arch of the 188-foot-long structure spans dry land to enclose cattle being herded along the creek. At one time this sleepy rural area teemed with industry: a massive, three-story mill and an ironworks in addition to the forge. Old Forge Farm, which encompasses a 1764 house, a barn shed, and a tenant house, stands east of the bridge. In 1893 the bridge was rebuilt by Josiah Hill.

U.S. Route 40 over Conococheague Creek

1936, Wilson, Washington County

➡ This triple-span, reinforced-concrete arch took over for Wilson Bridge in 1936, when heavy automobile and truck traffic became too much for the older bridge. The new bridge, an open-spandrel arch, is an elegant design that pays tribute to the old stone arch. The road, still called the National Pike, was chartered by Maryland in 1792 as a turnpike to connect Frederick and Cumberland. Opened in 1823, it was impassable by 1889, when a storm wrecked many of its bridges. The Good Roads Act of 1916 provided federal funding to repair the road, which has since been numerically designated as U.S. Route 40.

Blue Bridge · 1954, Cumberland, Allegany County

An unmistakable Cumberland landmark, the Blue Bridge carries Johnson Street (Maryland Route 942) over the North Branch of the Potomac River. This 315-foot-long, double-span bridge is one of the state's few steel tied-arch designs. A tied-arch bridge works much like a bow (as in a bow and arrow) turned on its side. Just as a bow forms an arch because it is tied together by a string, the arch of this bridge curves and is tied into place by long steel eyebars. The road hangs from the arch by metal suspenders. The spikes, here known as anti-climb shields, share the same purpose as those on the metal-arch Baltimore City bridges: to keep off would-be climbers. In the 1950s blue was a widely available paint color and a popular choice for bridges.

123

Washington County
Permanent Bridge
No. 15. Built by
JOHN A. WEAVER
for the Commissioners.
Jacob H. Eberie Pres.t
Andrew Rinehi.
Michael Smith.
Horatio N. Harne.
Samuel Lyday.
James Coudy.
Eli Orangetan.
Robert Coulter.
John C. Dorsey.
June the 24th
1839.

HISTORIC BRIDGES OF MARYLAND

In 1995 the Maryland State Highway Administration joined with the Maryland Historical Trust and the Federal Highway Administration to begin an inventory of Maryland's highway bridges. The initial phase of this inventory, part of the Maryland Inventory of Historic Properties, focused on bridges constructed between 1809 and 1947; the second phase, focusing on bridges constructed between 1948 and 1960, began in 2002. During the first phase, 855 bridges were identified as historic resources. Of these, 415 were determined to be eligible for inclusion in the National Register of Historic Places, based on the National Register's evaluation criteria. The following list includes these eligible bridges. Those marked with an asterisk [*] have been formally approved and are already listed in the National Register of Historic Places. Those marked with a dagger [†] are National Historic Landmarks.

ALLEGANY COUNTY

City/Town	Built	Name/Location	Type
Bellegrove	1925	U.S. 40 Scenic Route (McFarland Road) over Sideling Hill Creek	Concrete arch
Cumberland	1928	Market Street over Western Maryland Scenic Railroad rail lines	Metal girder
Cumberland	1932	U.S. 40 Alternate (National Pike) over Wills Creek	Concrete arch
Cumberland	1954	Blue Bridge: Md. 942 over North Branch of Potomac River	Metal arch
Flintstone	1900, 1925	Md. 144 (National Pike) over Flintstone Creek	Concrete arch
Flintstone	1919	Town Creek Road no. 3 over Town Creek	Concrete arch
Flintstone	1925	Md. 144 (National Pike) over Town Creek	Concrete arch
Frostburg	1915	Bowery Street over abandoned spur of Western Maryland Scenic Railroad rail lines	Concrete arch
Gildin	1910	Old Williams Road over Town Creek	Metal truss
Keifars	1932	Md. 51 over Chesapeake and Ohio Canal	Metal truss
Mill Run	1933	Md. 935 over George's Creek	Metal girder
Morrisons	ca. 1890–1900	Morrison Road over George's Creek	Metal truss
Mount Savage	1929	Md. 36 over Jennings Run	Concrete beam
Oldtown	1932	Md. 51 (Uhl Highway) over Mill Run	Concrete slab
Town Creek	1932	Md. 51 (Oldtown Road) over Town Creek	Metal truss
Town Creek	1932	Md. 51 (Uhl Highway) over Sawpit Run	Concrete slab

ANNE ARUNDEL COUNTY

City/Town	Built	Name/Location	Type
Annapolis	1946	Annapolis-Eastport Bridge: Md. 181 (Compromise Street) over Spa Creek	Movable
Davidsonville	1935	Md. 214 over Patuxent River	Metal truss
Elkridge	1944	O'Conner Road over Deep Run	Metal girder
Odenton/Ridgeway	1936	Md. 170 (Telegraph Road) over Severn Run	Concrete beam
Riviera Beach/Orchard	1946	Stoney Creek Bridge: Md. 173 (Fort Smallwood Road) over Stoney Creek	Movable
Severn	1931	Md. 174 over Amtrak rail lines	Metal girder

ANNE ARUNDEL/QUEEN ANNE'S COUNTIES

City/Town	Built	Name/Location	Type
Sandy Point	1949–52	William Preston Lane Jr. Memorial Bridge: U.S. 50/301 over Chesapeake Bay	Metal suspension
Sandy Point	1969–73	William Preston Lane Jr. Memorial Bridge: U.S. 50/301 over Chesapeake Bay	Metal suspension

BALTIMORE CITY

City/Town	Built	Name/Location	Type
Baltimore	1829	*†Carrollton Viaduct over Gwynns Falls	Masonry arch
Baltimore	1873	Fulton Avenue Bridge: Fulton Avenue over Amtrak rail lines	Masonry arch
Baltimore	1873	Vincent Street Bridge: Vincent Street over Amtrak rail lines	Masonry arch
Baltimore	1891–96	North Avenue Bridge: North Avenue over Falls Road, Jones Falls, Amtrak and CSX Railroad rail lines	Masonry arch
Baltimore	1895	Barclay Street Bridge: Barclay Street over CSX Railroad rail lines	Masonry arch
Baltimore	1895	Greenmount Avenue Bridge: Greenmount Avenue over CSX Railroad rail lines	Masonry arch
Baltimore	1895	Guilford Avenue Bridge: Guilford Avenue over CSX Railroad rail lines	Masonry arch
Baltimore	1895	Harford Road Bridge: Harford Road over CSX Railroad rail lines	Masonry arch
Baltimore	1895	Huntington Avenue Bridge: Huntington Avenue over CSX Railroad rail lines	Masonry arch
Baltimore	1900	Loch Raven Road over Minebank Run	Metal girder
Baltimore	1908	University Parkway over Stoney Run	Concrete arch
Baltimore	1911	Herring Run Bridge (Harford Road Bridge): Md. 147 (Harford Road) over Herring Run	Concrete arch
Baltimore	1916	Hanover Street Bridge: Hanover Street over Middle Branch of Patapsco River	Movable/concrete

City/Town	Built	Name/Location	Type
Baltimore	1919	Annapolis Road over Baltimore-Washington Parkway	Metal girder
Baltimore	1920	Fort Avenue over CSX Railroad rail lines	Metal girder
Baltimore	1920	Windsor Mill Road over Gwynns Falls	Concrete arch
Baltimore	1920	Wyman Park Drive over Stoney Run	Concrete arch
Baltimore	1921	Herring Run Bridge: U.S. 40 (Pulaski Highway) over Herring Run	Concrete arch
Baltimore	1925	Belair Road Bridge: U.S. 1 (Belair Road) over Herring Run	Concrete arch
Baltimore	1925	Frederick Road at Caton Avenue over Amtrak rail lines	Metal girder
Baltimore	1927	Clifton Avenue over Windsor Mill Road	Concrete arch
Baltimore	1928	Forest Park Avenue over Gwynns Falls	Concrete arch
Baltimore	1928	Mannasota Avenue over Herring Run	Concrete arch
Baltimore	1929	Cold Spring Lane over Herring Run	Concrete arch
Baltimore	1930	Frederick Avenue over Gwynns Falls and CSX Railroad rail lines	Concrete arch
Baltimore	1930	San Martin Drive over Stoney Run	Concrete arch
Baltimore	1931	Baltimore Street Bridge: Baltimore Street over Gwynns Falls and CSX Railroad rail lines	Concrete arch
Baltimore	1931	Lafayette Avenue over Amtrak rail lines	Metal girder
Baltimore	1932	Loch Raven Boulevard over Chinquapin Run	Concrete arch
Baltimore	1935	Sinclair Lane over CSX Railroad rail lines	Metal girder
Baltimore	1936, 1960	Guilford Avenue Bridge: Guilford Avenue over I-83 and Amtrak rail lines	Metal arch
Baltimore	1936	Orleans Street Viaduct: U.S. 40 over Jones Falls Expressway	Metal girder
Baltimore	1936	Wilkens Avenue over Gwynns Falls	Concrete arch
Baltimore	1937	29th Street Bridge: 29th Street over I-83 ramp, CSX and MTA rail lines, Falls Road	Concrete arch
Baltimore	1938	Hilton Parkway Bridge: Hilton Parkway over Franklintown Road and Gwynns Falls	Concrete arch
Baltimore	1938	Hilton Parkway over Gelston Run	Concrete arch
Baltimore	1938	Howard Street Bridge: Howard Street over I-83, Amtrak rail lines, and Jones Falls	Metal arch
Baltimore	1938	Mt. Royal Avenue over Howard Street	Concrete rigid frame
Baltimore	1940	Echodale Avenue over Herring Run	Concrete rigid frame

BALTIMORE COUNTY

City/Town	Built	Name/Location	Type
Arbutus	1931	Near Md. 166 over Baltimore and Ohio Railroad rail lines	Metal girder
Arcadia	1925	Dark Hollow Road over Little Piney Run	Concrete beam
Baldwin	1931	Md. 165 (Baldwin Mill Road) over Little Gunpowder Falls	Metal girder
Beckleysville	1931	Beckleysville Road over Pretty Boy Reservoir	Metal truss
Beckleysville	1932	Md. 25 (Falls Road) over George's Run	Concrete beam
Beckleysville	1933	George's Creek Road over George's Run	Concrete beam
Bentley Springs	1920	Eagle Mill Road over Little Falls	Concrete beam
Blenheim	1920	Dance Mill Road over Dulaney Valley Branch	Concrete beam
Blenheim	1930	Loch Raven Road over Dulaney Valley Run	Concrete beam
Bradshaw	1927	Little Gunpowder Bridge: Md. 7 (Philadelphia Road) over Little Gunpowder Falls	Concrete arch
Bradshaw	1935	U.S. 40 (Pulaski Highway) over Gunpowder Falls	Metal girder
Butler	1892–93	Cuba Road Bridge: Cuba Road over Western Run	Metal truss
Carney	1879	Loch Raven Road Bridge: Loch Raven Road over Towson Run	Masonry arch
Catonsville	1920	Thistle Road over unnamed stream	Concrete beam
Catonsville	1936	Patapsco River Bridge: U.S. 40 (National Pike, Edmondson Avenue Extension) over Patapsco River	Concrete arch
Chase	1916	Earls Road over Amtrak rail lines	Concrete beam
Cockeysville	1920	Sunnybrook Road over Greene Branch	Metal girder
Cockeysville	1922	Warren Road over Loch Raven Reservoir	Metal truss
Cockeysville	1924	Warren Road over Beaver Dam Run tributary	Concrete slab
Dover	1945	Md. 128 over Piney Run	Metal girder
Essex–Middle River	1929	Race Road over Stemmers Run	Concrete arch
Essex–Middle River	1933	Golden Ring Road over Stemmers Run	Concrete arch
Glyndon	1920	Tufton Avenue over Waterspout Run	Metal girder
Glyndon	1940	Belmont Avenue over Delaware Run	Metal girder
Glyndon	1947	Glyndon Bridge: Md. 128 (Butler Road) over Western Maryland Railroad rail lines	Concrete slab
Halethorpe	1936	U.S. 1 Alternate North (Washington Boulevard) over Amtrak rail lines and Herbert Run	Metal girder
Hebbville	1920	Ridge Road over Bens Run	Concrete beam
Hereford	1898	Masemore Road Bridge: Masemore Road over Gunpowder Falls	Metal truss

City/Town	Built	Name/Location	Type
Hereford	1924	Gunpowder Falls Bridge: Md. 45 (York Road) over Gunpowder Falls	Concrete arch
Hollofield	1929	Dogwood Road over Dogwood Run	Concrete arch
Hollofield	1934	Johnnycake Road over Patapsco River	Metal truss
Hunt Valley	1917	Md. 45 (York Road) over Western Run	Concrete arch
Hunt Valley	1922	Paper Mill Road Bridge: Paper Mill Road over Loch Raven Reservoir	Metal truss
Kingsville	1865, 1937	Jericho Road Covered Bridge: Jericho Road over Little Gunpowder Falls	Timber truss
Middle River	1942	Md. 150 over Md. 700	Metal girder
Mt. Vista	1915	Md. 147 (Harford Road) over Haystack Branch	Concrete arch
Mt. Vista	1915	Md. 147 (Harford Road) over Long Green Creek	Concrete arch
Owings Mills	1941	Md. 37 over Western Maryland Railroad rail lines	Concrete slab
Parkton	1809	Parkton Stone Arch Bridge: Md. 463 over Little Gunpowder Falls	Masonry arch
Parkton	1929	Md. 45 (York Road) over Little Gunpowder Falls	Concrete arch
Perry Hall	1924	Cromwells Bridge: Glen Arm Road over Gunpowder Falls	Concrete arch
Phoenix	1879	Carroll Road Bridge: Carroll Road over Carroll Branch	Metal truss
Pikesville	1907	Sudbrook Lane over CSX Railroad rail lines	Metal girder
Piney Grove	1907	Piney Grove Road over CSX Railroad rail lines	Timber
Pretty Boy	1932	Pretty Boy Dam Bridge: Pretty Boy Dam Road over Pretty Boy Reservoir	Concrete arch
Randallstown	1920	Offutt Road over Brice Run Branch	Concrete beam
Randallstown	1930	Old Court Road Bridge: Md. 125 (Old Court Road) over Brice Run	Concrete arch
Rayville	1933	Spook Hill Road over Frog Hollow Run	Concrete beam
Reckford	1928	Md. 147 (Harford Road) over Little Gunpowder Falls	Concrete arch
Reckord	1934	U.S. 1 over Little Gunpowder Falls	Metal girder
Rosedale	1934	U.S. 40 over Red House Creek	Concrete beam
St. James Corners	1920	Corbett Road over Carroll Branch	Concrete beam
Stingtown	1931	Stringtown Road over Black Rock Run	Concrete beam
Towson	1932	Md. 25 over Jones Falls	Concrete beam
Upperco	1907	Dover Road over CSX Railroad rail lines	Timber
White Hall	1933	Big Falls Road over Gunpowder Falls	Concrete arch
Whitemarsh	1935	U.S. 40 over Honeygo Run	Concrete beam
Whitemarsh	1935	U.S. 40 over Whitemarsh Run	Concrete beam
Woodlawn	ca. 1903	Md. 126 (Gwynn Oak Avenue) over Gwynns Falls	Concrete arch
Woodlawn	1928	Franklintown Road over Dead Run	Concrete arch

BALTIMORE/HARFORD COUNTIES

City/Town	Built	Name/Location	Type
Franklinville	1884	Vinegar Hill Road Bridge: Vinegar Hill Road (Franklinville Road) over Little Gunpowder Falls	Metal truss
Joppatowne	1935	U.S. 40 over Little Gunpowder Falls	Concrete beam

CALVERT COUNTY

City/Town	Built	Name/Location	Type
Chesapeake Beach	1940	Md. 261 over Fishing Creek	Metal girder

CALVERT/CHARLES COUNTIES

City/Town	Built	Name/Location	Type
Bowens/Benedict	1950	Benedict Bridge: Md. 231 over Patuxent River	Movable

CAROLINE COUNTY

City/Town	Built	Name/Location	Type
Bridgetown	ca. 1920	Md. 304 (Ruthsburg Road) over Long Marsh Ditch	Concrete arch
Denton	1911	Legion Road over Watts Creek	Concrete beam
Denton	1936	Hobbs Road (Md. 107F) over Watts Creek	Metal girder
Federalsburg	1910	Md. 315 (East Central Avenue) over Marshyhope Creek	Concrete arch
Greensboro	1932	Forge Branch Bridge: Md. 480 (Ridgely Road) over Forge Branch	Concrete arch
Goldsboro	1919	Sandy Island Bridge: Md. 287 (Sandtown Road) over Choptank River	Concrete arch
Hillsboro	1911	Tuckahoe Road over Tuckahoe Creek tributary	Concrete arch
Queen Anne	1915	Md. 404 Alternate over Tuckahoe Creek	Concrete beam
Three Corners	1919	Boyce Mill Road over Gravelly Branch	Concrete arch

CARROLL COUNTY

City/Town	Built	Name/Location	Type
Avondale	1936	Stone Chapel Road over Little Pipe Creek	Metal girder
Franklinville	1930	Md. 850 over Talbot Branch	Concrete slab
Lineboro	1929	Md. 86 over branch of Gunpowder Falls	Concrete slab
New Windsor	1908	Pearre Road Bridge: Pearre Road over Sams Creek	Metal truss
New Windsor	1924	Md. 31 over Dickerson Run	Concrete beam
New Windsor	1929	Md. 31 over Sams Creek	Concrete slab
Taneytown	1929	Md. 832 over Big Pipe Creek	Metal girder
Taneytown	1941	Md. 194 (Francis Scott Key Highway) over Big Pipe Creek	Metal girder
Union Mills	1934	Md. 97 over Big Pipe Creek	Concrete rigid frame

CECIL COUNTY

City/Town	Built	Name/Location	Type
Bayview	ca. 1860	Gilpin's Falls Covered Bridge over Northeast Creek	Timber
Chesapeake City	1948	Chesapeake City Bridge: Md. 213 over Chesapeake and Delaware Canal	Metal arch
Childs	1932	Md. 545 over Little Elk Creek	Metal truss
Conowingo	ca. 1885	Bell Manor Road Bridge: Bell Manor Road over Conowingo Creek	Metal truss
Elk Mills	1921	Appleton Road over CSX Railroad rail lines	Metal girder
Elkton	ca. 1936	Old Elk Neck Road (New Cut Road) over Piney Creek	Timber
Kilby Corner	1930	U.S. 222 over Octoraro Creek	Metal girder
New Bridge	ca. 1890	New Bridge Road over Octoraro Creek	Metal truss
North East	ca. 1885–1900	Rolling Mill Road Bridge: Rolling Mill Road over Northeast Creek	Metal truss
North East	1922	Northeast Creek Bridge: Md. 7 (Old Philadelphia Road) over Northeast Creek	Concrete arch
North East	1931	Md. 7 (Old Philadelphia Road) over Stoney Run	Concrete slab
North East	ca. 1940	Deaver Road over Baltimore and Ohio Railroad rail lines	Metal girder
North East	1944	Md. 272 over Northeast Creek	Concrete slab
Oakwood	ca. 1885–1900	McCauley Road Bridge: McCauley Road over Basin Run	Metal truss
Octoraro	1913	Horseshoe Road over Stone Creek	Concrete slab
Perryville	1940	U.S. 40 (Pulaski Highway) over Principio Creek	Concrete arch
Pleasant Hill	1930	Kirks Mill Lane over Northeast Creek	Metal girder
Richardsmere	1934	U.S. 1 over Octoraro Creek	Metal girder

CHARLES COUNTY

City/Town	Built	Name/Location	Type
Allens Fresh	1933	Md. 234 over Zekiah Swamp	Metal girder
Aquasco	1934	Asquasco Road over Swanson Creek	Concrete slab
Bryantown	1917, 1931	Md. 5 South (Leonardtown Road) over Zekiah Swamp	Concrete beam
Bryantown	1933	Bryantown Road over Mill Dam Run	Concrete beam
Chicamuxen	1928	Md. 224 over Reeders Run	Concrete beam
Grayton	1922	Md. 6 over Nanjemoy Creek	Concrete slab
Hawthorne	1929	Md. 225 over Port Tobacco Creek	Concrete beam
Mason Springs	1929	Md. 225 (Hawthorne Road) over branch of Mattawoman Creek	Concrete arch
Nanjemoy	ca. 1924	Liverpool Point Road over Beaverdam Creek	Concrete slab
Newburg	1939–49	Governor Harry W. Nice Memorial Bridge: U.S. 301 over Potomac River	Metal cantilever
Newport	1933	Stines Store Road over Gilbert Swamp Run	Concrete beam
Wayside	1920	Rock Point Road over Ditchley Prong	Concrete beam
Welcome	1929	Md. 6 (Port Tobacco Road) over Wards Run	Concrete arch

DORCHESTER COUNTY

City/Town	Built	Name/Location	Type
Brookview	1931	Brookview Bridge: Md. 14 over Marshyhope Creek	Movable
Cambridge	1939	Cambridge Bridge: Md. 795 over Cambridge Creek	Movable
Hare Town	1946	Bestpitch Ferry Road Bridge: Bestpitch Ferry Road over Transquaking River	Timber

City/Town	Built	Name/Location	Type
Bells Mill	1934	Catoctin River Bridge: Md. 464 (Olive School House Road) over Catoctin Creek	Metal truss
Braddock Heights	1936	U.S. 40 over Rock Creek	Concrete beam
Bridgeport	1925	Md. 140 (Taneytown Pike) over Monocacy River	Concrete arch
Brunswick	ca. 1910	Sumantown Road over Catoctin Creek	Metal truss
Brunswick	1925	Md. 478 over branch of Potomac River	Concrete beam
Buckeystown	1929	Md. 85 over branch of Monocacy River	Concrete slab
Burkittsville	1878	*Poffenberger Road Bridge: Poffenberger Road over Catoctin Creek	Metal truss
Burkittsville	ca. 1900–10	St. Marks Road over Broad Run	Metal truss
Catoctin	ca. 1872	Detour Road Bridge (now pedestrian bridge)	Metal truss
Catoctin Furnace	1927	Md. 806A over Little Hunting Creek	Concrete beam
Creagerstown	1914	Stevens Road over Hunting Creek	Metal truss
Dickerson	1931	Md. 28 over Monocacy River	Metal truss
Ellerton	1897	Crow Rock Bridge: Crow Rock Road over Middle Creek	Metal truss
Emmitsburg	ca. 1876	Four Points Bridge: Four Points–Keysville Road over Toms Creek	Metal truss
Emmitsburg	1908	Bullfrog Road Bridge: Bullfrog Road over Monocacy River	Metal truss
Emmitsburg	1914	Grimes Road over Toms Creek	Metal truss
Emmitsburg	1923	Toms Creek Bridge: U.S. 15 Business (Seton Avenue South) over Toms Creek	Concrete arch
Emmitsburg	1927	U.S. 15 Business (Catoctin Mountain Highway) over Flat Run	Concrete arch
Emmitsburg	1928	Creamery Road Bridge: Creamery Road over Toms Creek	Metal truss
Emmitsburg	1932	Md. 140 over Flat Run	Concrete beam
Emmitsburg	1932	Md. 140 over Middle Creek	Concrete beam
Four Points	ca. 1915	Sixes Bridge Road over Monocacy River	Metal truss
Frederick	1927	Md. 351 over Ballenger Creek	Concrete slab
Frederick	1930	Md. 355 (Urbana Pike) over Monocacy River	Metal truss
Frederick	1930	Yellow Springs Road over Little Tuscarora Creek	Concrete slab
Frederick	1931	Md. 26 (Liberty Road) over Israel Creek	Metal girder
Harmony	1918	Harmony Road over Little Catoctin Creek	Metal truss
Ijamsville	ca. 1910	Reichs Ford Road over Bush Creek	Metal truss
Johnsville	ca. 1890–1900	Simpson's Mill Road Bridge: Simpson's Mill Road over Little Pipe Creek	Metal truss
Knoxville	1920	Md. 180 over Potomac River tributary	Concrete beam
Knoxville	1926	Md. 478 over Potomac River tributary	Concrete beam
Lewistown	1931	Hessons Bridge Road over Fishing Creek	Metal girder
Libertytown	1929	Unionville Road over Weldon Creek	Concrete beam
Middlepoint	1932	Md. 17 over Middle Creek	Concrete beam
Middletown	ca. 1880	Bennies Hill Road Bridge: Bennies Hill Road over Catoctin Creek	Metal truss
Middletown	ca. 1920s	Holter Road over Hollow Creek	Concrete slab
Middletown	1920–25	Lewistown Road over Fishing Creek	Concrete beam
Middletown	1923	Catoctin Creek Bridge: U.S. 40 Alternate (National Pike) over Catoctin Creek	Concrete arch
Middletown	1934	Green Bridge: Md. 17 over Catoctin Creek	Metal truss
Myersville	ca. 1900	Mount Tabor Station Bridge: Station Road over Frostown Branch	Metal truss
Myersville	1919	Md. 17 over Little Catoctin Creek	Concrete beam
Myersville	1927	Md. 17 over Middle Creek	Concrete beam
Myersville	1928	Md. 17 over Catoctin Creek	Metal girder
Myersville	1930	Md. 17 (Wolfsville Road) over Middle Creek	Concrete beam
Myersville	1936	U.S. 40 (National Pike) over Catoctin Creek	Concrete arch
Myersville	1936	U.S. 40 (National Pike) over Little Catoctin Creek	Concrete arch
Myersville	1936	U.S. 40 (National Pike) over Middle Creek	Concrete arch
Myersville	1936	U.S. 40 (National Pike) over branch of Little Catoctin Creek	Concrete beam
Petersville	1912, 1932	Md. 180 (Jefferson Pike) over Little Catoctin Creek	Concrete beam
Petersville	1928	Md. 180 (Jefferson Pike) over Catoctin Creek	Concrete arch
Point of Rocks	1930	Md. 28 over Tuscarora Creek	Concrete beam
Point of Rocks	1937	Md. 28 over branch of Potomac River	Concrete slab
Point of Rocks	1939	U.S. 15 over Baltimore and Ohio Railroad rail lines and Potomac River	Metal truss
Rocky Ridge	1932	Md. 77 (Rocky Ridge Road) over Monocacy River	Metal truss
Rocky Ridge	1932	Md. 77 (Rocky Ridge Road) over Owens Creek	Metal girder

City/Town	Built	Name/Location	Type
Rosemount	1931, 1933	Md. 464 over Catoctin Creek	Concrete beam
Sabillasville	1936	Md. 550 over branch of Friends Creek	Concrete beam
Thurmont	ca. 1850	*Roddy Road Covered Bridge: Roddy Road over Owens Creek	Timber truss
Thurmont	1860	*Utica Mills Covered Bridge: Utica Road over Fishing Creek	Timber truss
Thurmont	1882	*Old Mill Road Bridge: Old Mill Road over Owens Creek	Metal truss
Thurmont	1887	Hoovers Mill Road over Owens Creek	Metal truss
Thurmont	1900	*Loys Station Covered Bridge: Old Frederick Road over Owens Creek	Timber truss
Thurmont	1914	Blacks Mill Road over Hunting Creek	Metal truss
Thurmont	1917	Apples Church Road Bridge: Apples Church Road over Owens Creek	Metal truss
Urbana	1904	Dixon Road over Bennett Creek	Metal truss
Urbana	1930	Md. 75 over branch of Bennett Creek	Concrete slab
Woodsboro	1900	*LeGore Bridge: LeGore Bridge Road over Monocacy River	Masonry arch

GARRETT COUNTY

City/Town	Built	Name/Location	Type
Altamont	1930	Md. 135 (Oakland-Westernport Road) over CSX Railroad rail lines	Metal girder
Asher Glade	1933	Md. 42 over Glade Run	Concrete beam
Avilton	1909	Avilton-Lonaconing Road over Savage River	Concrete arch
Bloomington	1937	Md. 135 over Savage River	Concrete rigid frame
Crellin	1921	Crellin Underwood Road over Snowy Creek	Concrete arch
Deer Park	1930	Fricks Crossing over Little Youghiogheny and CSX Railroad rail lines	Timber
Friendsville	1920	Old Selbysport Road (Cemetery Road) over Bear Creek	Concrete arch
Friendsville	1921	Accident-Friendsville Road over Bear Creek	Concrete arch
Friendsville	1921	Bear Creek Bridge (Accident–Bear Creek Bridge) over Bear Creek	Concrete arch
Friendsville	1932	Md. 828 over Youghiogheny River	Metal girder
Friendsville	ca. 1933	Md. 42 (Friendsville-Hoyes Road) over Buffalo Run	Concrete arch
Frostburg	ca. 1815	Old U.S. 40 over Little Savage River (off U.S. 40 Alternate)	Masonry arch
Grantsville	1813	*†Casselman River Bridge: National Road over Casselman River	Masonry arch
Grantsville	1926	Maple Grove Road over Casselman River	Concrete arch
Grantsville	1932	U.S. 40 Alternate over Casselman River	Metal truss
Jennings	1917	Jennings Road over South Branch of Casselman River	Concrete arch
Jennings	1931	Md. 495 over branch of Casselman River	Concrete slab
Merrill	1935	Savage River Road over Poplar Lick Run	Metal girder
Mineral Spring	1917	Chet Kelly Road over Mill Run	Concrete arch
Mineral Spring	1935	Mill Run Spur over Mill Run	Metal girder
New Germany	1935	Big Run Road over Big Run	Metal girder
Oakland	1913	Third Street over CSX Railroad rail lines	Trestle
Sand Spring	1919	Old Morgantown Road over Buffalo Run	Concrete arch

HARFORD COUNTY

City/Town	Built	Name/Location	Type
Aberdeen	1925	Md. 7 (Old Philadelphia Road) over Gray's Run	Concrete beam
Bel Air	1913	Lake Fanny Road over Winter's Run	Concrete arch
Bel Air	1930	Winter's Run Bridge: U.S. 1 Business (Bel Air Road) over Winter's Run	Concrete arch
Conowingo	1927	Conowingo Dam Bridge: U.S. 1 over Susquehanna River	Concrete beam
Creswell	1930	Md. 136 over James Run	Concrete beam
Dublin	1925	Forge Hill Road over Deer Creek	Concrete arch
Fallston	1930	Old Fallston Road over Maryland and Pennsylvania Railroad rail lines	Concrete slab
Five Forks	1930	Md. 136 over Falling Branch	Concrete beam
Forest Hill	ca. 1885–1900	Watervale Road over Winter's Run	Metal truss
Harkins	1934	Md. 136 over Big Branch	Concrete beam
Havre de Grace	1939–40	Thomas J. Hatem Memorial Bridge: U.S. 40 over Susquehanna River	Metal truss
Putnam	1931	Md. 165 over West Branch	Concrete beam
Pylesville	1928	Broad Creek Bridge: Old Pylesville Road over Broad Creek	Metal girder
Rocks	1931	Md. 165 over Little Deer Creek	Concrete arch
Rocks	1934	Md. 24 over Deer Creek	Metal truss
Rocks State Park	ca. 1885–1900	Cherry Hill Road over Deer Creek	Metal truss

City/Town	Built	Name/Location	Type
Sewell	1940	Abingdon Road over Baltimore and Ohio Railroad rail lines	Metal girder
Trappe	1883	Noble's Mill Bridge: Noble's Mill Road over Deer Creek	Metal truss
Trappe	1930	Priest Ford Road Bridge: Md. 136 over Deer Creek	Metal truss

HOWARD COUNTY

City/Town	Built	Name/Location	Type
Clarksville	1927	Sheppard Lane over Middle Patuxent River	Concrete arch
Columbia	ca. 1920	Old Columbia Road over Middle Patuxent River	Concrete arch
Dayton	ca. 1930	Triadelphia Mill Road over Patuxent River	Concrete slab
Dorsey	1937	Md. 176 over Deep Run	Concrete beam
Elkridge	1832–35	*†Thomas Viaduct: CSX Railroad rail lines over Patapsco River	Masonry arch
Elkridge	1940	River Road over Rockburn Branch	Metal girder
Ellicott City	1935	Tiber Alley over Tiber River	Metal girder
Ellicott City	1936	U.S. 40 over Forest Road Underpass	Concrete slab
Ellicott City	1939	U.S. 40 over Little Patuxent River	Concrete rigid frame
Font Hill Manor	1930	Frederick Road over Little Patuxent River	Concrete arch
Henryton	1925	Henryton Road over Patapsco River tributary	Concrete beam
Jessup	1936	Md. 32 over Baltimore and Ohio Railroad rail lines	Metal girder
Locust Ridge	1922	Middle Patuxent River Bridge: Folly Quarter Road over Middle Patuxent River	Concrete arch
Savage	1869	*†Bollman Truss Suspension Bridge over Little Patuxent River	Metal suspension
Sykesville	1962	Md. 32 over River Road, Patapsco River, and Baltimore and Ohio Railroad rail lines	Metal girder

KENT COUNTY

City/Town	Built	Name/Location	Type
Hopewell	1934	Hopewell Bridge: Md. 291 over Morgan Creek	Metal truss
Millington	1928	Md. 291 over Cypress Creek	Concrete beam
Sassafras	1913	Md. 299 (Sassafras Road) over Herring Branch of Sassafras River	Concrete arch
Sassafras	1938	Md. 299 (Massey Road) over Jacobs Creek	Timber and concrete

MONTGOMERY COUNTY

City/Town	Built	Name/Location	Type
Ashton	1928	Snell Bridge: Md. 108 (Ashton Road) over Patuxent River	Concrete arch
Ashton	1929	Md. 650 over Hawlings River	Concrete beam
Barnesville	1935	Barnesville Road (Md. 117) over Little Monocacy River	Concrete beam
Bethesda	1931	Md. 547 over Rock Creek	Concrete beam
Brighton	1941–44	Brighton Dam Road Bridge: Brighton Dam Road over Triadelphia Reservoir	Concrete slab
Brookeville	ca. 1920	Griffith Road over Hawlings River	Concrete arch
Brookeville	1927	Md. 97 over Reddy Branch	Concrete aeam
Bucklodge	1932	Md. 117 (Bucklodge Road) over Bucklodge Branch	Concrete slab
Dawsonville	1910	Montevideo Road over Dry Seneca Creek	Metal truss
Dickerson	1925	Md. 28 over Little Monocacy River	Concrete beam
Germantown	1905, 1929	Md. 118 over Baltimore and Ohio Railroad rail lines	Concrete beam
Glen Echo	1853–63	*†Cabin John Aqueduct: MacArthur Boulevard over Cabin John Creek and Parkway	Masonry arch
Poolesville	1925	Schaeffer Road over Little Seneca Creek	Metal girder
Poolesville	1931	Whites Ferry Road over Broad Run	Concrete slab
Rockville	1911	Baltimore Road over Rock Creek tributary	Concrete arch
Sellman	1928	Peach Tree Road over CSX Railroad rail lines	Metal girder
Silver Spring	1918	Talbot Avenue over CSX Railroad rail lines	Metal girder
Silver Spring	1931	Park Valley Road over Sligo Creek	Concrete beam
Sunshine	1930	Md. 97 over Hawlings River	Concrete beam
Takoma Park	1932	Sligo Creek Bridge: Md. 195 (Carroll Avenue) over Sligo Creek	Concrete arch

PRINCE GEORGE'S COUNTY

City/Town	Built	Name/Location	Type
Bowie	ca. 1907–10	Governor's Bridge: Governor's Bridge Road over Patuxent River	Metal truss
Bowie	ca. 1920	Md. 978 over Collington Branch	Concrete slab
Bowie	1926	Md. 450 over Conrail rail lines	Metal girder

City/Town	Built	Name/Location	Type
Croom	ca. 1930	Md. 382 over Mataponi Creek	Concrete slab
Croom	1933	Md. 382 over Charles Branch	Concrete beam
Fort Washington Forest	1932	Livingston Road over Piscataway Creek	Concrete beam
Greenbelt	1927	Md. 201 (Edmonston Road) over Beaverdam Creek	Concrete arch
Greenbelt	1937	Md. 212 over Baltimore and Ohio Railroad rail lines	Metal girder
Greenbelt	1937	Md. 212 over Indian Creek	Concrete slab
Riverdale	1931	Md. 412 Alternate (Riverdale Road) over Northeast Branch	Concrete arch
Suitland	ca. 1910	Wheeler Hill Road over Barnaby Run	Concrete slab
Takoma Park	1931	Northwest Branch Bridge: Md. 212 (Riggs Road) over Northwest Branch	Concrete arch
Takoma Park	1934	Md. 410 over Sligo Creek	Concrete beam
Upper Marlboro	1928	Md. 725 over Federal Spring Branch	Concrete beam

QUEEN ANNE'S COUNTY

City/Town	Built	Name/Location	Type
Centerville	1934	Md. 213 over Gravel Run	Concrete slab
Centreville	1945	Md. 213 over Old Mill Stream Branch	Concrete slab
Church Hill	1911	Md. 19 Alternate over Southeast Creek	Concrete beam
Church Hill	1933	Md. 405 over Southeast Creek	Concrete beam
Price	1931	Md. 405 (Price Station Road) over German Branch	Concrete slab
Queenstown	1929	Md. 456 over branch of Wye River	Concrete slab
Ruthsburg	1915	Md. 304 over German Branch	Concrete slab

SOMERSET COUNTY

City/Town	Built	Name/Location	Type
Pocomoke City	1933	Md. 364 over Dividing Creek	Metal girder

ST. MARY'S COUNTY

City/Town	Built	Name/Location	Type
Chingville	1938	Md. 244 (Beauvue Road) over Poplar Hill Creek	Metal girder
Cremona	1930	Md. 6 (Turner Road) over Persimmon Creek	Concrete slab
Great Mills	1932	Md. 471 over St. Mary's River	Concrete beam
Huntersville	1930	Md. 6 (Turner Road) over Lockes Swamp Creek	Concrete slab
Maddox	1929	Md. 238 (Maddox Road) over Burroughs Run	Concrete srch

TALBOT COUNTY

City/Town	Built	Name/Location	Type
Easton	1932	Dover Bridge: Md. 331 over Choptank River	Movable/metal truss
Longwoods	1911	Md. 662 over Potts Mill Creek	Concrete slab
Queen Anne	1928	Md. 303 (Lewistown Road) over Norwich Creek	Concrete slab

WASHINGTON COUNTY

City/Town	Built	Name/Location	Type
Antietam	1832	Antietam Iron Works Bridge: Harpers Ferry Road over Antietam Creek	Masonry arch
Antietam Nat'l Battlefield	1836	Burnside Bridge over Antietam Creek	Masonry arch
Benevola	ca. 1820s	Kline's Mill Bridge (Newcomer's Mill Bridge): U.S. 40 Alternate over Beaver Creek	Masonry arch
Benevola	ca. 1920	Old Roxbury Road over Beaver Creek	Concrete slab
Big Pool	1938	U.S. 40 over Licking Creek	Metal truss
Boonsboro	ca. 1824	Devil's Backbone Bridge: Md. 68 over Beaver Creek	Masonry arch
Boonsboro	1833	Booth's Mill Bridge (Delemere Bridge): Md. 68 over Antietam Creek	Masonry arch
Boonsboro	1906	Barnes Road over Beaver Creek	Concrete arch
Boonsboro	ca. 1922	Md. 858 (Appletown Road) over Dog Creek tributary	Concrete slab
Cearfoss	1829	Broadfording Road Bridge: Broadfording Road over Conococheague Creek	Masonry arch
Clear Spring	1907	Md. 56 (Big Pool Road) over Little Conococheague Creek	Concrete arch
Fairview	1932	Conococheague Creek Bridge: Md. 494 over Conococheague Creek	Metal truss
Funkstown	1823	Funkstown Turnpike Bridge: U.S. 40 Alternate over Antietam Creek	Masonry arch
Funkstown	1824	Roxbury Mill Bridge: Roxbury Road over Antietam Creek	Masonry arch
Funkstown	1833	Shafer's Mill Bridge (Funkstown Bridge): East Oak Ridge Drive over Antietam Creek	Masonry arch

City/Town	Built	Name/Location	Type
Funkstown	1839	Rose's Mill Bridge: Garis Shop Road over Antietam Creek	Masonry arch
Funkstown	1840	Claggett's Mill Bridge: Poffenberger Road over Antietam Creek	Masonry arch
Funkstown	ca. 1841	Claggett's Mill Race Bridge: Poffenberger Road over Antietam Creek	Masonry arch
Gapland	ca. 1922	Gapland Road over Israel Creek	Concrete slab
Hagerstown	1863, 1893	Old Forge Road Bridge: Old Forge Road over Antietam Creek	Masonry arch
Hagerstown	ca. 1900	Prospect Street over Antietam Street	Metal girder
Hagerstown	1934	Antietam Creek Bridge: Md. 64 (Jefferson Boulevard) over Antietam Creek	Concrete arch
Hagerstown	1936	U.S. 40 over Landis Spring Branch	Concrete rigid frame
Hagerstown	1936	U.S. 40 West (National Pike) over Antietam Creek	Concrete arch
Hagerstown	1941	U.S. 40 over branch of Antietam Creek	Concrete rigid frame
Hancock	1916	Old U.S. 40 East (Tollgate Ridge Road) over Tonoloway Creek	Concrete arch
Hancock	1937	U.S. 522 over Md. 144 and Tonoloway Creek	Metal girder
Hancock	1937–39	U.S. 522 over Potomac River, CSX Railroad rail lines, and Chesapeake and Ohio Canal	Metal truss
Keedysville	1830	Hitt Bridge (Upper Bridge): Keedysville Road over Antietam Creek	Masonry arch
Keedysville	1832	Hess Mill Bridge: Coffman Lane over Little Antietam Creek	Masonry arch
Keedysville	after 1835	Pry's Mill Bridge: Keedysville Road over Little Antietam Creek	Masonry arch
Keedysville	ca. 1850	Felfoot Bridge: Dogstreet Road over Little Antietam Creek	Masonry arch
Keedysville	1927	Little Antietam Creek Bridge: Md. 845 A (Main Street) over Little Antietam Creek	Concrete arch
Leitersburg	1839	Strite's Mill Bridge: Leiters Mill Road over Little Antietam Creek	Masonry arch
Leitersbrug	1908	Clopper Road Bridge: Clopper Road over Antietam Creek	Concrete arch
Leitersburg	1931	Md. 62 over Little Antietam Creek	Concrete slab
Locust Grove	ca. 1922	Marble Quarry Road over Little Antietam Creek	Concrete slab
Rohrersville	1922, ca. 1930	Md. 858 (Main Street) over Little Antietam Creek	Concrete slab
Sandy Hook	1946–47	Sandy Hook Bridge: U.S. 340 over Potomac River, CSX Railroad lines, Chesapeake and Ohio Canal	Metal truss
Sharpsburg	1937–39	James Rumsey Bridge: Md. 34/W.Va. 480 over Potomac River	Metal truss
Smithsburg	1915, 1955	Wolfesville Road over Beaver Creek	Concrete beam
Wagners Crossroads	1936	U.S. 40 over Beaver Creek	Concrete beam
Wagners Crossroads	1936	U.S. 40 over Little Beaver Creek	Concrete rigid frame
Williamsport	1829	Md. 68 over Conococheague Creek	Masonry arch
Williamsport	ca. 1935	Cedar Ridge Road Bridge: Cedar Ridge Road over Meadow Brook Creek	Masonry arch
Wilson	1817–19	*Wilson Bridge over Conococheague Creek (now pedestrian bridge)	Masonry arch
Wilson	1936	U.S. 40 (National Pike) over Conococheague Creek	Concrete arch

WICOMICO COUNTY

City/Town	Built	Name/Location	Type
Avalon	1940	U.S. 13 North over Leonard's Mill Pond Run	Concrete beam
Pittsville	1934	Md. 353 over Burnt Mill Branch	Concrete slab
Pittsville	ca. 1939	Warren Road over Campbell's Ditch	Timber
Quantico	1926	Md. 347 (Quantico Road) over Quantico Creek	Concrete slab
Salisbury	1927	Wicomico River Bridge: Md. 991 over Wicomico River	Movable
Salisbury	1928	Snow Hill Road (Route 12) over East Branch of Wicomico River	Concrete beam
Salisbury	1930	U.S. 13 Business over Baltimore and Ohio Railroad rail lines	Metal girder
Salisbury	1937	U.S. 13 Business over East Branch of Wicomico River	Timber and concrete
Whaleysville	1946	Md. 346 (Ocean City Road) over Pocomoke River	Concrete beam

WORCESTER COUNTY

City/Town	Built	Name/Location	Type
Ocean City	1942	Ocean City Bridge: U.S. 50 over Sinepuxent River	Movable
Ocean City	1942	U.S. 50 West over Herring Creek	Concrete beam
Pocomoke City	1920	Pocomoke City Bridge: Md. 675 over Pocomoke River	Movable
Pocomoke City	1945	U.S. 13 South over Wagram Creek	Timber and concrete
Snow Hill	ca. 1932	U.S. 113 South over Corkers Creek	Concrete slab
Snow Hill	1932	Snow Hill Bridge: Md. 12 over Pocomoke River	Movable
Whiton	1932	Md. 354 over Tilghman Race	Concrete beam

ACKNOWLEDGMENTS

The author wishes to thank the following individuals for their assistance: Jill Dowling; Anne Bruder, Valerie Burnette-Edgar, Liz Buxton, Jock Freedman, Bruce Grey, John Hudacek, Becky Kermes, Neil Pedersen, Lora Rakowski, Doug Simmons, Cynthia Simpson, Don Sparklin, Kelley Steele, Rita Suffness, and Matt Zulkowski, Maryland State Highway Administration; Dan Johnson and Joe Policelli, Federal Highway Administration; Mary Louise deSarran, Nicole Diehlmann, and Elizabeth Hughes, Maryland Historical Trust; Abba Lichtenstein, bridge consultant; John W. McGrain, Baltimore County Planning Department; Janet Davis, Frederick County Planning Department; Mike Dixon and Joan Mihich, Cecil County Historical Society; Cameron Chasten and Ed Voigt, U.S. Army Corps of Engineers; Ross Kimmel and John Purgason, Maryland Department of Natural Resources; Kenneth C. Harwood Jr., Frederick County Public Works Department; Nick Deros and C. Michael Dougherty, URS Corporation; Richard Chen, chief, Bridge Engineering, Baltimore City; Terry McGee, Washington County Engineering Department; Elizabeth Graff, Washington County Historical Society; Jane Hershey and Pat Schooley, Washington County historians; John Frye, Washington County Free Library; John Howard, National Park Service; and Eric DeLony, Historic American Engineering Record, National Park Service. A special thanks goes to Diane Maddex, Gretchen Smith Mui, Robert L. Wiser, John Legler, and Pedro E. Guerrero.

PHOTOGRAPHS

Credits

All photographs by Carol M. Highsmith except for the following:

New-York Historical Society: 100 (negative number 74984)

The Hughes Company, Courtesy Maryland State Highway Administration: 4, 8 (all), 10–11, 19 top, 23 top, 28, 68, back endpaper

Marion E. Warren: 6

Maryland Historical Society, Baltimore: front endpaper, 2, 16, 19 bottom, 20, 21, 23 bottom, 50

Display Photographs

Case binding: William Preston Lane Jr. Memorial Bridge (Chesapeake Bay Bridge) (1949–52, 1969–73), Sandy Point, Anne Arundel and Queen Anne's Counties. Front endpaper: Carrollton Viaduct (1829), Baltimore City. Page 1: Guilford Avenue Bridge (1936), Baltimore City. Pages 2 and 3: Burnside Bridge (1836), Antietam National Battlefield, Washington County. Page 4: Deer Creek Bridge (Maryland Route 161) (1930, not extant), Darlington, Harford County. Page 5: Crow Rock Bridge (Crow Rock Road) (1897), Ellerton, Frederick County. Pages 6 and 7: William Preston Lane Jr. Memorial Bridge (Chesapeake Bay Bridge) (1949–52, 1969–73), Sandy Point, Anne Arundel and Queen Anne's Counties. Pages 8 and 9: Snow Hill Bridge (1932), Snow Hill, Worcester County. Page 24 top: Pry's Mill Bridge, ca. 1835, Keedysville, Washington County. Page 24 center: Loys Station Covered Bridge, 1880, Thurmont, Frederick County. Page 24 bottom: Pretty Boy Dam Bridge, 1932, Pretty Boy, Baltimore County. Page 25 top: Rolling Mill Road Bridge, ca. 1885, North East, Cecil County. Page 25 center: Blue Bridge, 1954, Cumberland, Allegany County. Page 25 bottom: Cambridge Bridge, 1939, Cambridge, Dorchester County. Page 124: Plaque commemorating the bridge builder John A. Wever, Rose's Mill Bridge (1839), Funkstown, Washington County. Back endpaper: Kitzmiller Bridge (late 1800s, not extant), Kitzmiller, Garrett County.

Bullfrog Road Bridge
1908, Emmitsburg, Frederick County

INDEX